LOST WAX BRONZE CASTING

LOST WAX BRONZE CASTING

A photographic essay on this antique and venerable art

by HARRY JACKSON

Foreword by John Walker
DIRECTOR EMERITUS, NATIONAL GALLERY OF ART

NORTHLAND PRESS/FLAGSTAFF

Historical Research by Richard Fremantle

©Copyright 1972 by
Wyoming Foundry Studios Establishment

International Standard Book Number: 0-87358-093-1
Library of Congress Catalog Card Number: 72-188289

ALL RIGHTS RESERVED — PRINTED IN UNITED STATES OF AMERICA
NORTHLAND PRESS • FLAGSTAFF, ARIZONA 86001

To the Honorable Robert Coe

CONTENTS

Foreword	ix
Introduction	1
The Process	
1: THE CREATION AND PRESERVATION OF THE ORIGINAL SCULPTURE IN SOLID WAX	12
2: MAKING THE REPRODUCTORY WAX	22
3: RETOUCHING THE HOLLOW REPRODUCTORY WAX	40
4: EXTERIOR PREPARATION • TIE RODS, CIRCULATORY SYSTEM, OUTER INVESTMENT	46
5: THE BURN-OUT, WHERE THE WAX IS LOST	60
6: PREPARATION OF THE BURNED-OUT MOLDS AND POURING OF THE MOLTEN BRONZE	68
7: CLEANING AND ROUGH FINISHING THE BRONZE	78
8: CHASING AND FINAL FINISHING OF DETAILS	90
The Patination and the Painting of Bronzes	103
Glossary of Technical Terms	113
Selected Bibliography	123
Acknowledgments and Credits	127

FOREWORD

HOW OFTEN IN MY LONG CAREER as a curator and museum director have I been asked to explain *cire perdue*, or the lost wax method of bronze casting! And how often have I described it, always somewhat dubiously, and as it turns out never quite accurately! When Harry Jackson's manuscript was sent to me I was delighted to read at last an authoritative description of this technique. It is a much more complicated procedure than I had dreamt, but this book elucidates it in the only practical way — with photographs and long captions documenting every step in the fabrication of a piece of sculpture. To make a successful cast is a dramatic story, and I found it utterly absorbing.

I was so fascinated that I went back to the source of my knowledge of *cire perdue*, the *Autobiography of Benvenuto Cellini*. The methods have not changed, but the materials have improved. Cellini did not have the advantage of "a fine dental investment" to paint on the reproductory wax, or of "cold-setting synthetic rubber" for the master mold. But lack of modern paraphernalia does not explain the hair-raising difficulties he had casting the Perseus. There was an explosion of gases, which Harry Jackson explains can

Model for equestrian statue of Louis XIV, c. 1685–90, by Martin Desjardin. The Royal Museum of Fine Arts, Copenhagen.

easily happen, his studio caught fire, his bronze refused to flow. He had to throw into the molten metal all the pewter in his house. In the end he succeeded, but, as he says, it was a miracle. Unlike his Renaissance predecessor Harry Jackson does not tell us of his disasters, nor does he indicate that supernatural genius is required to overcome them. His account is straightforward and objective, but the difficulty of making a perfect cast, whether in the sixteenth century or the twentieth century, becomes obvious.

For this reason after reading Cellini and Jackson one realizes how important it is that the cast be made, or at least approved, by the sculptor himself. Then, when the bronze leaves the foundry, it has his imprimatur. Today this significant fact is generally disregarded. There is a museum in Paris which offers made-to-order bronzes by all the best-known recent French sculptors from Rodin to Despiau. Such posthumous casts present a problem. Even when the work of art is not cast by the sculptor, as Harry Jackson points out, a great deal of work remains to be done, especially in its patination.

What would Rodin and the other artists think of these replicas, which are now entering important collections? Should museums limit themselves to sculpture done during the lifetime of the artist and bearing his stamp of approval, or should they include casts made after his death? No self-respecting curator of prints, for example, would think of exhibiting an etching by Rembrandt pulled in the nineteenth century nor would he show an engraving by Piranesi printed today from his steel engraved plates. Are the standards of curators of prints higher than those of curators of sculpture? If so it is to be hoped that Harry Jackson's book, which should be on the desk of every sculpture curator, will bring about a better understanding and appreciation of quality in bronzes; and that someday collectors will show in the plastic arts as much discrimination in selecting casts as they do in the graphic arts in acquiring fine impressions.

JOHN WALKER
Director Emeritus
National Gallery of Art

INTRODUCTION

THE LOST WAX METHOD of casting bronze is at least 6,000 years old. It was used in most of the ancient cultures, and without it the bronze masterpieces of Greece, Rome, Egypt, and the Near East could not have been created. It has always been employed for casting gold and silver as well. Today's lost wax method is used to produce nearly all bronze sculpture and is based on the same ancient principles. Though it is an art which has been passed from father to son for some 6,000 years, almost nothing has been written or diagrammed to describe the steps and basic method.

Someone is always asking me to explain the method used for casting my sculpture. It can be done but, because it is so complex, it is usually more confounding than helpful. I always wanted a good clear book to recommend to my questioners, but there wasn't one. Like most people, I assumed that libraries had to have something on their shelves about a technique so deeply rooted in our cultural history, but after looking everywhere I found none existed. The misinformation regarding the making of bronze sculptures is shockingly widespread, not only among laymen but among art critics and historians as well.

Bull's head, cast by the lost wax method, from Sin Temple IX at Khafáje, c. 2500 B.C. Courtesy of the Oriental Institute, University of Chicago.

Offering stand from Temple Oval III at Khafáje, 2800–2600 B.C. Courtesy of the Oriental Institute, University of Chicago.

I would like to digress for a moment in order to illustrate this lack of general knowledge. Often my questioners, assuming that bronze is carved directly from the metal itself, will ask how one manages to achieve such action and detail in so hard and resistant a material. A few of my readers may be surprised at the misconception such a question implies but they ought not be too quick to scoff. In 1969, the English sculptor Henry Moore told me of a very famous art critic who complimented him one day by saying, "While standing before your most recent bronze, I felt such tremendous force and vigor in your direct audacious carving of the raw material itself." The great sculptor was too polite to correct this "art expert" who should have known that all bronzes are cast, and most of them by the lost wax method. Moore confirmed my findings that no book existed which dealt satisfactorily with the entire process. He agreed with the need for such a book and advised me to forge ahead and complete the present work.

I had already been encouraged to pursue this book project by two other renowned sculptors, Jacques Lipchitz and Paul Manship. Manship died in 1966, immediately after completing his heroic seventeen-foot figure of Teddy Roosevelt for the Theodore Roosevelt Island in the Potomac, Washington, D.C. Lipchitz at this writing is finishing a forty-foot sculpture of a winged Pegasus for Columbia University. These contemporary monuments can be immortalized in bronze only through the practice of the very ancient lost wax method.

Museum experts also confirmed the lack of any reliable comprehensive text on this subject. To wait 6,000 years is to wait too long. I decided to do a book that would begin to fill this need by describing as completely as possible one aspect of this vast subject, namely, the actual method of casting. The fascinating history of lost wax has been neglected almost as much as the technique, and though it is impossible to separate them completely, I have kept the historical references to a minimum in order to concentrate on the central theme of this book. To do justice to its history would require volumes and that is well outside my scope and intent.

It is astounding and most puzzling that this lack of technical information regarding lost wax has continued, despite its age-old and worldwide employment and the fact that today at least ninety per cent of all bronze sculpture is cast by some form of this method. The first prehistoric artists who discovered the basic principle of lost wax casting, *wherever there was wax, there will be bronze*, were only able to cast small crude, simple forms in solid metal. Later man learned to cast hollow bronzes of gigantic size and infinite refinement, but the same basic principle, *wherever there was wax, there will be bronze*, still held true and holds true today.

Most likely, with this rule in mind, those early primitive sculptors started with natural beeswax and created a small figure of a man, a horse, or an idol. When it was finished, they added wax extensions to the top and the bottom of the figure. Next, they completely enveloped the figure and extensions with a very thick coating of fine, liquid clay. But the ends of the wax extensions protruded slightly from this clay mold to form the simplest possible gate and vent system. This mold was heated in a fire until it was hard as brick and the wax figure and the extensions inside were melted. They were burned out. They were gone forever — lost. Wax was replaced with air.

Bronze was melted, and poured into the cavity left by the lost wax. It was poured through the upper extension, called the gate, forcing the air out through the lower extension, called a vent. When the bronze cooled, the baked clay was chipped off, leaving the rough bronze casting. When the gate and vent extensions were removed, the small figure was smoothed with ancient versions of today's chisels, rasps and hammers. This was the first version of the lost wax method.

North Syrian female figure, c. 4000 B.C. This was found among a cache of six statuettes on the Amuq Plain, by far the earliest objects cast in human form discovered to date. Height 5½". Courtesy of the Oriental Institute, University of Chicago.

Provincial horse from Peloponnesus, c. 1200–800 B.C. In the early history of Greek art, the Peloponnesian cities were noted for their fine bronze work. Staatliche Museen, West Berlin.

Male portrait head from Ife, Nigeria, A.D. 1100–1200. Designed for ancestor worship, the holes in this piece were used for inserting hair, possibly taken from the person represented. Height 13½". Collection, the Oni of Ife.

WARRIOR ON HORSEBACK *by the Italian Renaissance sculptor, Riccio. Victoria and Albert Museum, London.*

These early works were solid. The next major discovery occurred about 4,000 years ago during the Akkadian period in Mesopotamia. This was the method of casting bronzes with hollow interiors and only then did it become possible to cast larger works.

The French phrase for lost wax, *cire perdue,* is the most universally used term. To give an idea of the enduring prestige of the lost wax process, the Latin word, *sincerus,* meaning "without wax" originally applied to a lost wax bronze casting without a single imperfection. *Sincerus* is the root of our word, sincere, which to this day is defined as "real and true, without hypocrisy, embellishment, or exaggeration."

The story of lost wax began before history. I would like to touch on just a few outstanding examples and periods: small, solid statues from the Amuq Plain and objects found in a cave near Masada on the Dead Sea, about 6,000 years old; Chinese bronzes from the tomb of Huang-Ti, the Yellow Emperor, almost 5,000 years; bronze statues from Mesopotamia in the Akkadian period, about 4,000 years, and in India and Afghanistan, bronze fragments have been found that are also about 4,000 years old. The great bronzes from Buddhist southern India start about A.D. 500, and the African bronzes from Ife and Benin, Nigeria range from A.D. 1100 to 1600. The above bronzes and all those from ancient Egypt, Greece, Rome, and Etruria were cast by the lost wax method. In addition, many archaeologists believe that the method was used in the Americas at least as early as A.D. 1200 — that only it could account for the fine detail and complex form of the Incas' earliest figurative jewelry, cast in pure gold.

Written evidence of early bronze casting also goes far back in history. It is mentioned and described at great length in the Bible. A notable example is found in 1 Kings, where one of the early bronze artisans, Hiram of Tyre, is introduced: "A craftsman in bronze, wise, adroit and skillful at doing brazier's work; and to do such work King Solomon had now summoned him." He made "All the appurtenances of the Lord's Temple . . . such a multitude that he did not reckon the weight of all the bronze he used."

A treatise on bronze was written at the beginning of the Chou dynasty in 1122 B.C. An early Chinese manuscript, *K'ao kung chi*, written sometime after the first century A.D. gives copper-tin proportions for various bronze objects. A "wax process" for bronze casting is mentioned in South Indian manuscripts from the second century A.D. In 1125 A.D. an illustrated catalogue, *Hsüan-ho-po-ku t'u lu*, was compiled on bronzes in the Chinese imperial collection, some of them dating from the earliest periods of their bronze casting.

It was the Greeks, however, beginning in the sixth century B.C., who brought bronze casting to a perfection never reached before, except by the Chinese. But where the Chinese concentrated mainly on ceremonial vessels, and the Indians and Egyptians on symbolic images of their Gods, the Greeks chose to glorify man himself as the focal point of all nature. To achieve this the Greek sculptors turned more and more to large freestanding sculptures and to the use of bronze. Marble was static: it was necessary to "cut away" to carve it, and this, plus its natural fragility put significant limitations on what the artist could envision and execute. To make a bronze, however, a full-scale model was first created out of clay, wax, or other soft modeling material, easily worked, formed by the will of the artist, and perfect for the expression of the power and freedom of man.

Judging by their extreme thickness, the early Greek bronzes were probably cast directly in sand. Later Greek bronzes of the Roman period were much thinner. However, parts of even the earliest statues were cast by the lost wax method — in the case of the famous "Charioteer" from Delphi (475–470 B.C.) the head, arms, and feet were cast by lost wax. The skirt, drapery, and upper torso were cast in sand. The various parts then had to be joined and it is easy to see where the joints were made and the imperfections have been filled with bronze patches. This pro-

Head of an Akkadian ruler, c. 2250 B.C. Found almost perfectly preserved in a rubbish heap in Ninevah, it may be a portrait of the great ruler, Sargon. During his reign bronze work was already of the finest craftsmanship. Height 13⅝". The Iraq Museum, Baghdad.

Chinese Ting Vessel, B style, Yin or early Chou, c. 1100 B.C. Height 8½". Pillsbury Collection, the Minneapolis Institute of Arts.

PARVATI, *portrait of a queen from Indian Asia, c. A.D. 1000–1100. The excellent quality of early Indian bronzes indicates a long history of casting, undoubtedly originating in their frequent contacts with the Assryians, Greeks, and Romans. N. J. Treasurgwalla Collection. National Museum, New Delhi.*

cedure was common right through the great Roman period — 300 B.C. to A.D. 300.

In the fifth century B.C. the Greeks had solved most of the problems of bronze alloys and casting and by the third century had progressed to doing giant hollow bronze statues, which is verified by the writings of Pliny: "But calling for admiration before all others was the colossal Statue of the Sun at Rhodes made by Chares of Lindus, the pupil of Lysippus. . . . This Statue was 105 feet high; and 66 years after its erection (ca. 226 B.C.) was overthrown by an earthquake, but even lying on the ground it was a marvel. Few people can make their arms meet round the thumb of the figure, and the fingers are larger than most statues; and where the limbs have been broken off enormous cavities yawn. . . ."

Despite this evidence that the lost wax method developed as a continuous chain, linking pre-history to the present day, many people erroneously believe that the term "lost wax" described an old method, lost through disuse. But by now it should be clear that the term is derived from the process, itself, wherein the wax is lost. Though its use in the eastern world never ceased, it is true that it was "lost" to the West temporarily. The period of its oblivion in Europe corresponds with the Dark Ages, approximately A.D. 400 to A.D. 1100, when most of the classic techniques of painting and sculpture were forbidden and forgotten.

Slowly, through the Crusades and the consequent reopening of trade routes, the craft of lost wax casting was recovered from the Middle East and returned to Europe through the port cities of Venice and Palermo. From there it eventually took root again over the entire western world.

Soon after the year A.D. 1000 masterpieces cast by lost wax appeared: the doors of Saint Michael now in the cathedral of Hildesheim, the doors of the church of San Zeno in Verona, the Duomo in Pisa and those in Ravello and Monreale.

Bringing bronze artisans from Venice, Andrea da Pontedera completed the first set of doors for the Baptistery of the Duomo in Florence in 1338. By the beginning of the Renaissance in the early 1400s lost wax bronze casting had again become as important as it had been in ancient times. The competition for the second set of Baptistery doors held in 1401 read like a "Who's Who in World Sculpture" and included Brunelleschi, Jacopo della Quercia and Lorenzo Ghiberti. Ghiberti won, and completed them in 1424. He finished the third, or main doors, in 1452. These final doors were so admired by Michelangelo that he said, "They are worthy of being the true Gates of Paradise." The fine detail and the absolutely faithful reproduction of Ghiberti's original model are due to the unique qualities of lost wax casting.

From that time on it was used by all the great sculptors ranging from Donatello (1386–1466) to Rodin (1840–1917), our own Remington (1861–1909) and Russell (1864–1926), and our contemporaries Lipchitz and Moore on up to this very moment. Leonardo da Vinci (1452–1519) describes and diagrams various aspects, Benvenuto Cellini (1500–1571) goes into great detail in his famous autobiography but he isolates and brings forth only a few of its phases, while Giorgio Vasari (1511–1574) writes a too short thumbnail sketch in his "Techniques of the Artists." He ends his comments with glowing praise for the process:

"But that is a truly marvelous thing which is come to pass in our times, this mode of casting figures, large as well as small . . . tufts of rue and any other slender herb or flower can be cast in silver and in gold, quite easily and with such a success, that they are as beautiful as the natural; from which it is seen that this art is more excellent now than it was in the time of the ancients."

Bronzes from other civilizations, including those of the Renaissance and today, have a wall thickness of 3/16" to 3/4" or more.

The famous CHARIOTEER *from the Sanctuary of Apollo at Delphi, c. 470 B.C. Reins held in his hand illustrate perfectly the section on non-cast metal details in* PHASE XXX *of the text. Height 71". National Museum, Delphi.*

Young jockey done in Greece's Golden Age, 500–400 B.C. Though the horse he was seated on has been lost, the jockey expresses the speed and movement of riding, a good early example of the scope given the sculptor through the lost wax method. National Museum, Athens.

It is possible that the Romans surpassed all civilizations in this aspect of the art by producing bronzes which were only 1/16" to 1/8" thick. I've experienced the thrill of looking at a life-size Roman portrait bust which appeared to weigh a ton. Yet, when I picked it up, it was as light as an empty egg shell. The detail was perfect.

This thinness of the wall is extremely important in large castings because metal shrinks considerably as it cools, making large, solid works impossible to cast. Shrinkage distorts the shape, and ruins the sculpture's surface. With a thin wall of metal, rather than a solid mass, less shrinkage occurs, thus the thinner the wall, the more faithful the reproduction of the artist's original wax figure. Therefore the artist, holding in mind the basic principle of *wherever there was wax there will be bronze* makes his reproductory wax of as thin a shell of wax as possible rather than out of a solid chunk. The cavity inside the wax shell is then filled with a liquid clay which hardens to form a fire-resistant core. When the thin shell of wax is *lost* in the burn-out oven, and the molten bronze is poured into the empty space, the result is a perfect bronze replica of what had started as a thin wax shell. The closely related third development required enlarging and refining the circulatory system of the gates and vents. Thus the molten bronze could flow evenly and unhampered to all parts of the thin cavity left by the loss of even the most complex wax shells.

There were several early techniques used to make the fire-resistant core. The Greeks, among others, began with the core. In fact, it was their original sculpture. They made it out of clay and powdered firebrick. When it was dry, they covered it with a thin coat of wax, and then they added all of the detail they wanted by incising lines and designs, or by adding subtle accents with bits of wax stuck to the surface. All of these final touches were reproduced in bronze with great accuracy.

The next discovery was a technique for making a series of castings from the same original sculpture through the development of a master mold. In the beginning it was made of terra-cotta, and later of plaster. Its purpose was to capture a permanent negative impression of the original so that several reproductory wax shells could be formed. Remains of terra-cotta and plaster master molds which date back to at least 1000 B.C. have been found in Asia Minor. Greek examples date from 800 B.C., and those used by the Romans go back to 300 B.C. Yet I use a basically similar process today.

There's only one other way to make bronzes, and that is by sand casting. But this limits the sculpture to the most simple forms while lost wax can be used to produce works with unlimited intricacy. It frees the artist to create anything he desires in bronze. One of the best examples I know of is the bronze figure of a young jockey, done in Greece's Golden

Frederic Remington's THE OUTLAW, *1906. Remington took every advantage of the freedom allowed by the lost wax method, stretching it to its ultimate limits, as illustrated by the exquisite balance of this piece. Height 22¼". Courtesy of the Kennedy Galleries, Inc., New York.*

Age, 500–400 B.C. The horse isn't there anymore, but the jockey is, and to see him is to feel the charging horse beneath him. One forgets that he is sitting on air. He isn't. He's just sitting on an invisible horse.

In order to make it easier to follow and understand the process, I've used one sculpture to illustrate the lost wax method in this book. Everything shown was done in my studio, workshop-foundry in Camaiore, Italy. I created the original in wax, and my artisans and I have reproduced it several times in bronze. It is a piece called "Pony Express" and I chose it because of its complex form and successful balance. Because the entire piece is supported on the point of a single horsehoof, it puts the lost wax method of casting to the greatest possible test.

I hope this book succeeds in clarifying this exciting and still flourishing method. In 6,000 years we haven't found a better one for giving the sculptor creative freedom.

HARRY JACKSON
Wyoming Foundry Studios
Camaiore, Lucca, Italy

THE PROCESS

Original action study cast in bronze.

1: THE CREATION AND PRESERVATION OF THE ORIGINAL SCULPTURE IN SOLID WAX

PHASE 1: *Working from studies*

1. The artist executes an exact scale drawing of the final sculpture to be created in solid wax. He refers to two smaller, rough studies made in solid wax to capture the overall movement, mass, and drama.

2. The two rough wax sketches and finished full scale drawing.

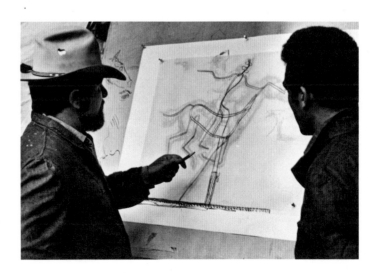

3. The artist makes a schematic drawing on tracing paper to guide the craftsman in making the metal supporting armature.

PHASE II: *The armature*

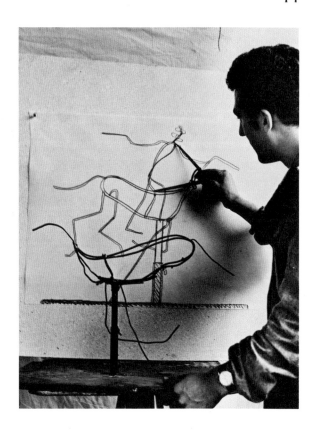

4. A pair of calipers is used to transfer the exact measurements to the metal armature under construction. It will support the heavy solid wax torso of the horse and rider. The unfinished armature is on a turntable in front of the drawing.

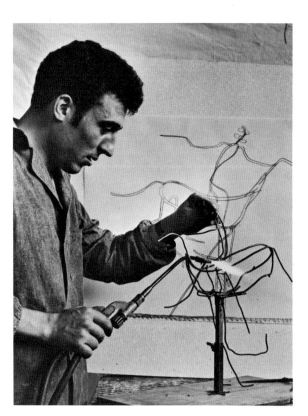

5. An acetylene torch and silver solder are used to weld the armature.

6. Close-up of welding torch and silver solder rod.

7. The finished armature.

PHASE III: *The original*

8. The artist takes a double handful of wax out of a pot of warm water. The water is warmed over a Bunsen burner (visible below the pot). Next to the burner is a glass filled with olive oil. The artist keeps his hands well greased with oil so that the wax does not adhere to his skin and become unmanageable. The wax is kneaded in its warm state until it gets to a good working consistency. On the right the armature can be seen with the first application of the modeling wax beginning to form the mass of the horse's torso.

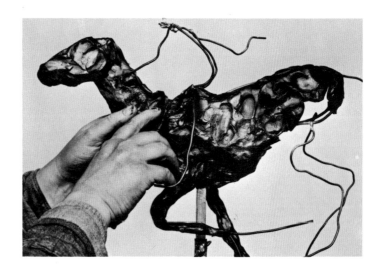

9. Close-up showing rough application of the wax mass.

10. Horse's head begins to take shape.

11. The artist uses a hot iron, which has been heated on the burner, to model some details on the horse's head. The rider is still quite undefined, and small gage armature wires stick out of the two hind legs to act as additional support during the modeling process.

12. The artist adds final touches to the right hind leg.

13. The original wax is almost completed.

14. A close-up of the rider's head.

PHASE IV: *The master mold*

The original model in this instance has been made of wax although a sculptor may create his model in any material which seems fitting and exciting. If the original work is made of a material that can be burned out and disappear (for example: wax, wood, weeds, grasses, or a modern synthetic substance such as styrofoam), then this original can be sacrificed if the artist chooses. It may be covered directly with the fireclay investment, after first having been provided with a *circulatory system*. (See Phase XVII.) If, on the other hand, the artist wishes to save his original or if he prefers to work in unburnable materials such as water clay or terra-cotta, plasticine, or plaster, then a negative mold must be made of the original. Possibilities are, for example, a plaster piece mold, a gelatin mold, or a mold of such substances as moulage (composed of a type of seaweed that forms a gelatinous mass at low temperature). This or a glue mold are good for only one or two wax impressions. A recent contribution to the field of bronze casting is the synthetic elastic mold, such as rubber Tuffy used for the bronze depicted in this book. A cold-setting latex, it eliminates the dangers of melting or damaging the original wax, and yet produces a mold which can make many more wax impressions than gelatin or other methods. The elastic mold material is applied to the wax original, and in turn is encased in a plaster outer shell or *couche*. The rubber mold, protected and held rigidly in place by its outer shell, makes the first negative impression of the "critical surface" of the artist's original. This delicate process requires the work of skilled artisans who achieve a faithful impression while protecting the original wax from damage.

15. Here is the original wax, in perfect condition and permanently preserved in a glass case.

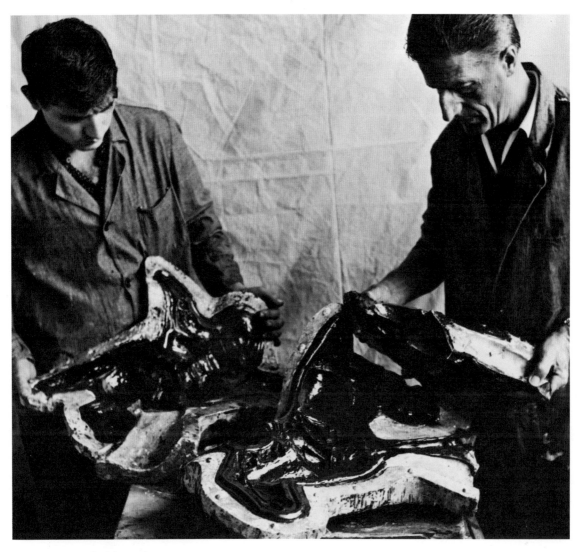

16. Artisans holding the open finished master mold. Inside the sections of the plaster shell lie the inner rubber negatives, shiny dark in the photograph.

2: MAKING THE REPRODUCTORY WAX

PHASE V: *Preparation*

17. The master mold is prepared in the wax shop.

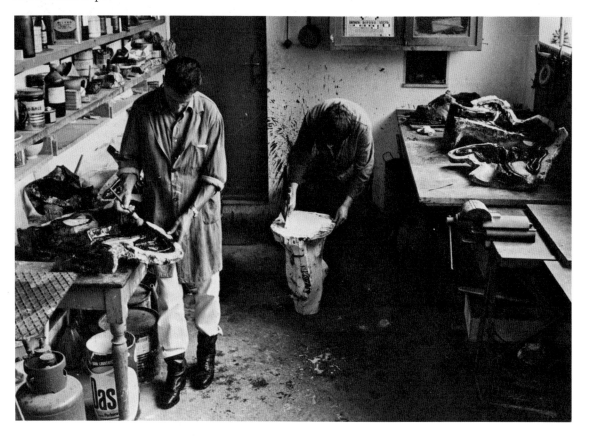

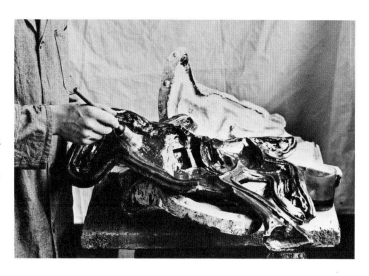

18. The rubber negative is loosely laid over the plaster shell while a very light film of liquid silicon is brushed onto the negative surface. This film will enable the subsequent application of the burnable wax positive to separate easily from the rubber negative. Silicon may also be sprayed on.

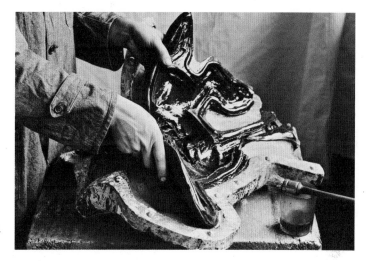

19. The prepared rubber negative is placed in the mother mold, or plaster *couche*, which is precisely formed to hold it and keep it rigid, while the molten wax is applied with a brush onto the negative surface.

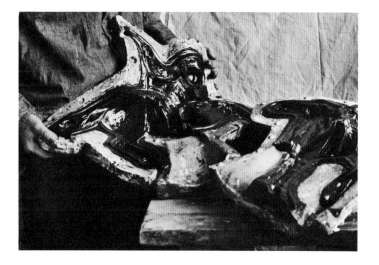

20. The two major halves of the negative master mold exposed.

PHASE VI: *The making of the wax alloy.*

21. Paraffin is melted into the warmed bucket. Gum resin, or *pece greca* as it is called in Italian, is next added to the mix. In the foreground is half a cake of pure beeswax, unrefined. It constitutes roughly one-half to three-quarters of the final wax composition and will be added last. Finally a small amount of nondrying oil is added to the mix, more or less depending upon the heat or cold of the season in which the work is to be done.

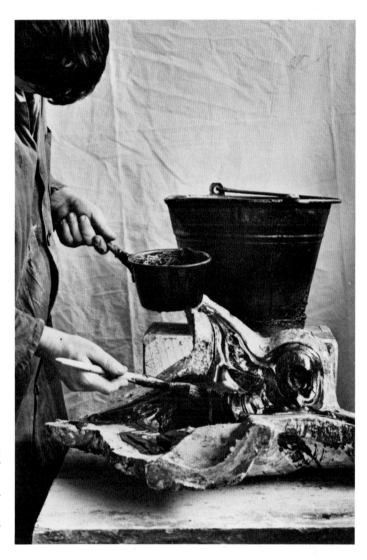

22. Molten wax is transferred to a small pot from which it is brushed onto the negative critical surface of the master rubber mold. The artisan makes sure that the liquid wax fills the most minute recesses and details of this negative rubber surface.

23. Close-up of the wax being brushed onto the surface. This thin veil of wax is not intended to create the final required thickness of the finished wax. That is accomplished in Phase VIII.

PHASE VII: *Locking the master mold*

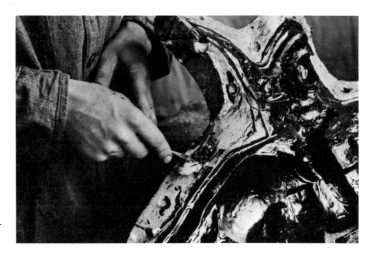

24. The brushing has been completed and the artisan, with a small metal tool, is cleaning the surface of the conjoining lip of one half of the plaster *couche* or "mother mold." He removes all the wax drippings which have inadvertently spilled during the brushing.

25. The two lips of the converging plaster surfaces have been cleaned perfectly. The locking negative and positive keys of the rubber mold must be cleaned as well, so that when they are joined there will be no extra wax or foreign substances to keep them from making a perfect, close-fitting joint.

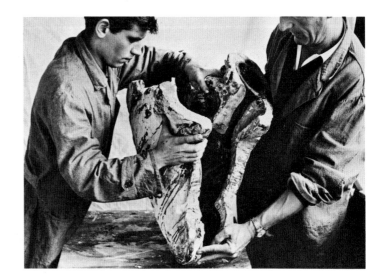

26. The master mold is fitted together.

27. Metal horseshoe clamps hold the plaster halves of the outer shell together in a rigid position for pouring the reproductory wax and fireclay core.

PHASE VIII: *Slushing*

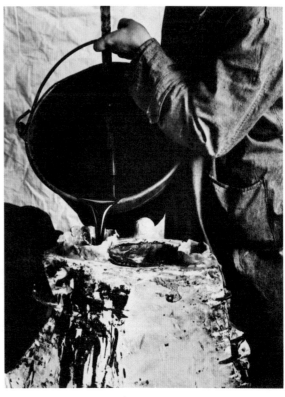

28. Molten wax is now poured into the hollow interior of the wax-coated mold and permitted to stand for a short time while it begins to harden from the outside in, automatically forming a very even shell which augments the thickness of the previously brushed on wax film.

29. This step is called "slushing" and is a most delicate and crucial process, as the thickness of the wax shell determines the final bronze's thickness, recalling the age-old rule of thumb: "Wherever there was wax there will be bronze." The still fluid wax is slushed around in the mold until sufficient wax adheres to the negative surface and has hardened. The entire mold is then reversed and the excess liquid wax poured back into the bucket.

30. Cold water is poured into the interior of the now thickened wax shell to help it harden and solidify quickly.

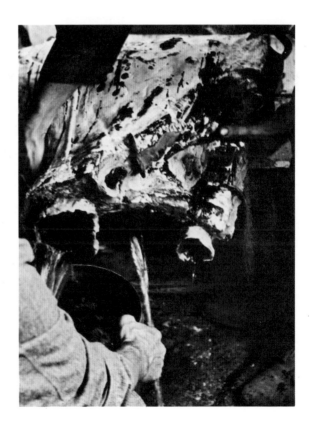

31. The water is poured out after the short time required to harden the wax.

After you have filled it with wax let it thoroughly cool for a whole day — if it be summer, say two days. . . . THE TREATISES OF BENVENUTO CELLINI ON GOLDSMITHING AND SCULPTURE.

In Cellini's time apparently they did not use the technique of cooling wax with water, but the seasonal considerations are still valid in almost every phase.

PHASE IX: *The breather tube*

The hollow interior of the hardened wax shell will be filled with a liquid mixture, essentially of fireclay. This fireclay interior investment is called the "core" and is all important, as it enables the finished bronze to be hollow. It must be made of a fire-resistant, rigid, and porous mass, which will crumble away easily for removal after the bronze has been cast. When the bronze is poured into the mold (Phase XXII) a great amount of gases will be formed in a split second by the intense heat and chemical transformation taking place. The escape of these gases must be controlled or they will cause major damage and possibly an explosion. The fireclay core is purposely made porous to permit the gases to reach and escape through the breathing tube, which is inserted into the hollow before the fireclay is introduced, and passes through the entire interior of the bronze. It is open at both ends and along its entire length to allow the gases ready escape.

32. A piece of galvanized iron sheeting is formed into a hollow tube, open along its entire length.

33. Close-up of this process.

34. The prepared tube has been bent in the proper curve to enable it to pass through the hollow interior and come out at the right point. The tube is about to be coated with molten wax so that fireclay will not clog the metal tube and block the gases from their controlled escape.

35. The wax-coated tube is being inserted into the hollow interior of the wax casting. A nail driven through one end of the tube serves as a spanner to keep it away from the edge of the wax casting. When the tube finally is in place, the spanner will keep it centered in the middle of the fireclay investment (to be poured in the next phase).

PHASE X: *Making the core investment*

36. Here we see the ingredients of the investment material which will be used to make the core. The bucket contains water; the scoop on the right contains plaster of paris; the larger pile in the left foreground is ground up firebrick, and the material being poured into the bucket is fine poplar sawdust. First the sawdust is put in and permitted to soak up well; next the fireclay, and finally the plaster. The sawdust burns out when the mold is finally put into an oven (in Phase XIX) and thus permits the inner core investment to be extremely porous, but because of the fireclay and plaster, quite rigid. It will be thoroughly honeycombed when the sawdust particles burn out under the continued slow fire applied to the mold in the oven.

37. A drill chuck with a mixing arm attachment is used to agitate and homogenize the core material.

38. The homogeneous material is poured slowly into the hollow form which contains the wax shell, closely laid against and reproducing the negative rubber mold. The wax-coated breathing tube is visible next to the stream of fireclay.

39. The last bits of fireclay core are poured into the interior. The breather tube is still seen extending out of the core.

40. The remaining investment, now quite a bit more glutinous and self-sustaining, is used to build up a base that extends beyond the actual bottom, which is upside down in this phase. When set back on its bottom, this hardened extension of fireclay will serve as a base upon which the piece will stand for retouching and preparation.

41. A piece of glass is used to smooth and level the surface to form a solid steady base.

PHASE XI: *Unlocking.* After about twenty-four hours the fireclay investment in the interior, or the core of the wax, has had sufficient time to harden and become self-supporting and rigid.

42. The form has been turned right side up again. The horseshoe grips are loosened with a hammer.

43. Close-up of the grip being broken loose.

44. The joints of the two major pieces are opened very slowly and delicately with large metal spatulas.

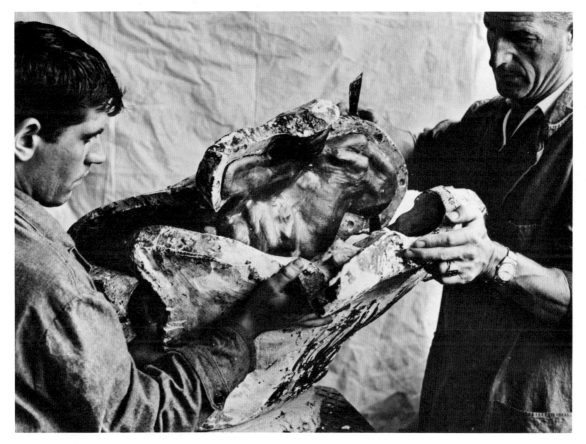

45. The plaster mother mold is being removed from one side, leaving the outside of the rubber negative mold visible.

46. The fine elastic negative mold is taken off to reveal the wax positive casting.

47. The rubber mold is carefully moved away from the small detail of the revolver here. The convex locking channel is clearly shown around the outer extremity of the rubber mold. It fits into a concave channel in the opposite section and helps lock the two halves of the master mold together.

48. The rubber mold is stripped off the final side, revealing the full wax casting. Also visible is the flashing (sometimes called fins or seams) formed by the wax seeping through the joints at the parting line of the rubber mold.

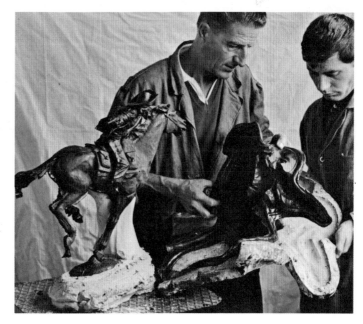

49. The rubber mold, now removed from the hollow wax casting, is carefully returned to its plaster retaining mold where it will remain until it is used again to make a second wax casting.

3: RETOUCHING THE HOLLOW REPRODUCTORY WAX

PHASE XII: *Preparatory retouching*

50. These are some of the tools required for retouching and working directly in the wax. The sculptor is holding a bit of retouching wax in his right hand and a metal tool in his left. On the edge of the modeling stand are metal rasps, knives, scalpels, and wooden tools—various shapes are needed to work the wax from difficult angles. The glass to the left is filled with kerosene which is used as a cutting and smoothing material because it melts and softens the wax. In the darker glass is olive oil, cut with a bit of kerosene to help lubricate the hands and the wooden cold cutting and modeling instruments to keep them from sticking to the wax.

51. An artisan uses a hot iron for retouching and removing the flashings that remain on the wax casting. This wax is removed either by cold cutting with a sharp instrument or, in some cases, by heating a blunt tool on the burner seen at the lower right. This photo clearly shows the fireclay base and the large column of fireclay going up just inside the sustaining leg of the horse. The fireclay support contains the bottom end of the main breather tube and acts as the primary support for the hollow wax and fireclay core while the work of retouching proceeds.

52. With a sharp tool the artisan cuts and defines the edges of a hole in the horse's head. He is exposing the fireclay core and searching for the breather tube, illustrated in Phase IX. The breather tube will be exposed, and an extension will be added so that it can eventually connect with the system of vents permitting the air and gases to escape the exterior mold. These various functions of cutting the breather holes, baring the breather tubes, and taking off the flashing are done by expert artisans because even this work, as simple as it may seem, has to be done with a very intmiate knowledge of the entire process.

PHASE XIII: *Final retouching*

53. The wax casting has been rough-cleaned, and now the sculptor begins the crucial retouching. Here he is warming a tool on the gas burner, just before going to the wax and changing and refining the surface. Because of the sculptor's final interpretive work, each wax is unique.

54. Close-up of the sculptor heating a tool on the burner. A device sculptors sometimes use is to hold a bit of wax in the mouth, like chewing gum, taking just a bit of it out when it is warm and workable. The emulsion of saliva is not in any way detrimental to the wax; in fact, it helps. It acts as a kind of isolating material to the hand and is superior to oil as it is not slippery.

55. The sculptor touches up the details and freshens the expression on the rider's face with a very fine sharp tool.

56. Close-up of the sculptor finishing the rider's face.

57. The palm of the sculptor's hand is moistened with olive oil and kerosene. The mixture is different for each person, depending upon how dry or oily one's natural skin is or how the day and the humidity affect the material. These are things that can never be formulated, and one just has to feel his way. Here the sculptor is rubbing the tip of the stick into the oil, after which he will model the wax. The wooden stick can also be heated very slightly over the fire or in the mouth and then rubbed in the oil. All of these little touches are vital in achieving the desired sculptured nuance.

58. The sculptor refining the modeling of the horse's eye.
"... *if you are minded to add any subtle labour or fancy to your work, you are easily able to do it.*"
THE TREATISES OF
BENVENUTO CELLINI ON GOLDSMITHING AND SCULPTURE.

4: EXTERIOR PREPARATION—TIE RODS, CIRCULATORY SYSTEM, OUTER INVESTMENT

PHASE XIV: *The tie rods*

59. Pictured are the nails and pins to be nailed into the wax casting through the critical surface. One container holds brass pins, exactly like the common straight pins used for sewing. In the other are brass nails. An artisan drives in these pins, which serve the same purpose as tie rods in pouring cement. In this instance, they will penetrate into the core material halfway and stick out into the air halfway. They are so placed that when the outer investment material is added and the wax burned out, leaving a completely empty space, the fireclay core is supported by these nails and pins and does not collapse against the sides of the outer fireclay investment, but maintains the desired distance. Then this empty space with its critical negative surface is left clear to receive the molten bronze and leave its impression on the critical positive surface of the bronze.

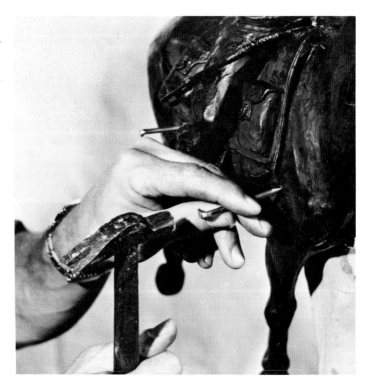

60. One of the nails being placed. Above the artisan's hand is another nail already in position.

61. The finished wax casting with all of the nails and pins in place, looking much like a prickly pear.

PHASE XV: *Additional core investment*
In this particular phase, the lower legs of the rider are about to be filled with fireclay core material.

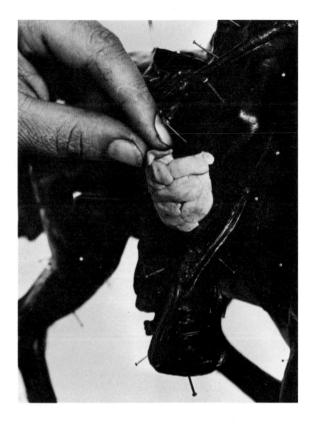

62. With a bit of light-colored wax a small funnel is built around the entrance hole, and here we see skilled hands forming the temporary funnel.

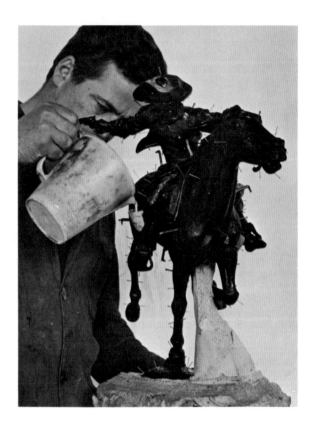

63. The liquid core material is poured in very slowly so that no air bubbles are formed and the air being displaced can come out evenly through the same hole. It's a bit like pouring water into a small-necked bottle.

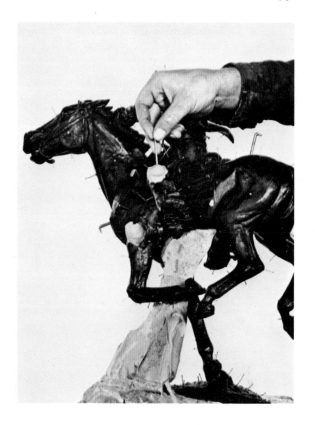

64. When the legs are completely filled, small breather tubes are lowered to the bottom of the rider's legs. Their tops protrude to permit attachment to the final venting system.

65. A detail of two of the funnels and breather tubes. Later we will see the hole left in the bronze surface by this procedure. I have added this particular phase so the reader will understand why these holes exist in the rough bronze casting. They are not faults but a very necessary part of the core-venting of a lost wax casting.

PHASE XVI: *Gating.* The circulatory system has a complex nomenclature. In England it is different from that in America. In Italy all passages leading to the cavity are called *collate* and all those leading away are called *respiratori*. In American terminology the sprues, runners, and gates permit the metal to go into the cavity, and the breathing tubes, risers, and vents permit air to escape.

66. Breather tube extensions are seen sticking out of the rider's hat and the forehead of the horse. We also see the two main arteries or *sprues* for the *runners and gates;* they are standing upright on either side of the horse. The main *sprues* go off on their smaller capillary extensions (*runners*) which, by the way, always go up from the *sprue* canal because it increases the pressure of the metal and decreases the chance of air blockage. These small arterial *runners* go from the main canal (*sprue*) over to the critical surface of the wax casting. The sprues are made, in this case, of an Italian light cane, the sort used in cane chairs. It is also very commonly done with wax tubes that are prepared beforehand in various gages and dimensions. Many other burnable materials can be employed, but in all cases wax is used for joining them to each other and to the casting.

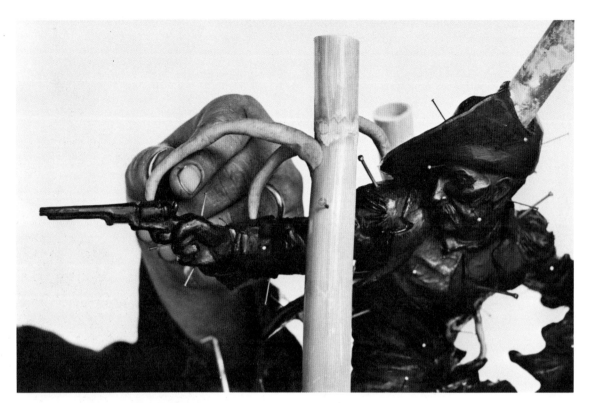

67. A close-up of one of the capillary extensions or runners of the sprue being attached to the cylinder of the revolver. Another one in the background touches the wrist of the man's glove. Again, in the upper right-hand part we see the extension of the breather tube as it issues from the hole made in the upturned brim of the rider's hat. All of the extremities of the piece have to be vented. By venting, we mean the placement of an air escape or breather tube.

68, 69, and 70. We see the fully gated wax casting from three sides, ready to be covered with the outer fireclay investment. Visible at the top is a plastic funnel which will form the mouth into which the bronze will eventually be poured. We also see clearly the very thin vents or breather tubes rising high in the air. In Photo 70 the wax tubes hanging down from the horse's right hoof, his chin, and his tail are very important traps to catch the minute amount of waste material which is inevitably left over; theoretically it should not exist, but in practice it always does. A little dirt or residue material left over from the burned out cane and wax will go down into those traps where it does no harm to the surface of the final piece.

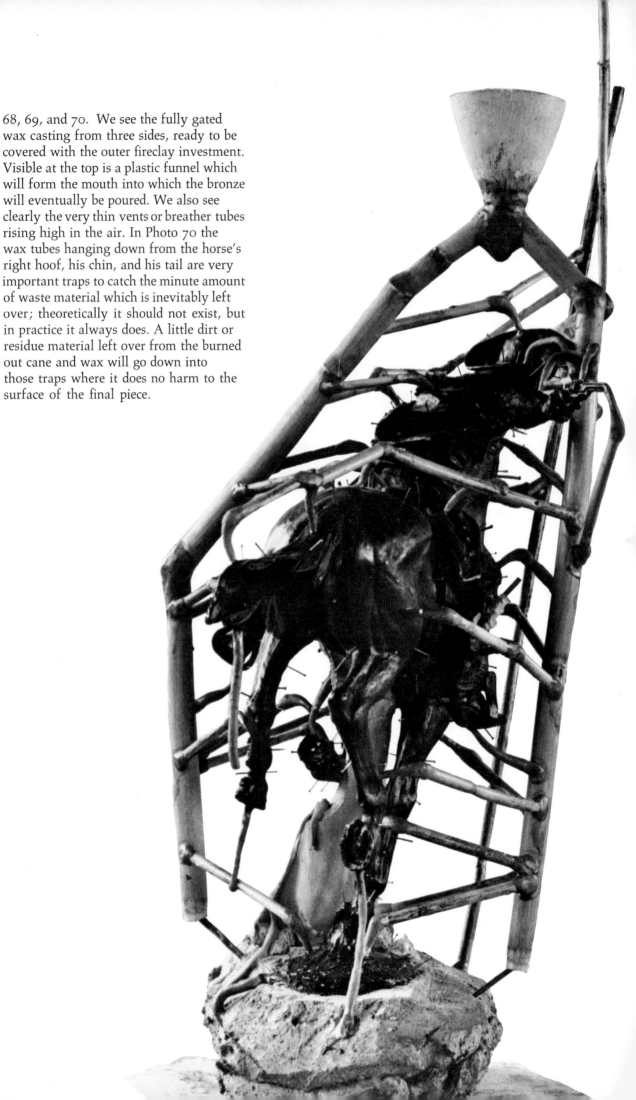

54

PHASE XVII: *First outer investment*

71. The first layer of the outer investment material in a very liquid form is being carefully brushed onto the surface. This first layer, which will constitute one of the negative critical surfaces, is made of a very fine material, called dental investment, purchased from a dental supply house. Because of its stable consistency, it does not shrink or deform in any way,
thus giving a perfect impression. This expensive and delicate material is used for the first few millimeters of the critical surface. If the layer is applied too thinly, it will not be resistant and will
separate from the outer rough coat, made of the usual fireclay.

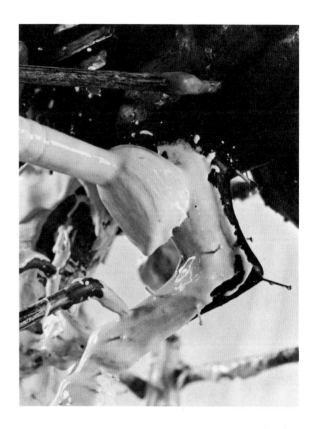

72. A detail of the dental investment being applied by brush. Naturally, no air holes can be allowed to form. If they do, they will fill up with bronze and reproduce themselves positively as small bumps on the surface.

73. Here the wax is about three-quarters covered. Notice that the sprues and vents are not intentionally covered. It makes no difference, really, whether they are covered or not. It is only the critical surface that must be so carefully covered.

PHASE XVIII: *Outer mold*

74. In this photograph the piece is completely covered with the dental investment, and we see the beginnings of the base of the outer mold, done in the rougher fireclay. In contrast to the core fireclay, the outer mold must not be either fragile or collapsible. Therefore, sawdust is excluded from its composition. The gases are of less concern in this outer mold than in the core mold, for here they would pass more freely into the open air, and not be entrapped within the bronze casting.

75. This outer mold wall is built of a more solid mixture of fireclay so that it is rigid and self-sustaining. It is built up almost the way some primitive Indian might build up a very thick pot, and serves as a circular retaining wall.

76. Now it is almost completed. Next to it is a pail of the same material, but thinned with water.

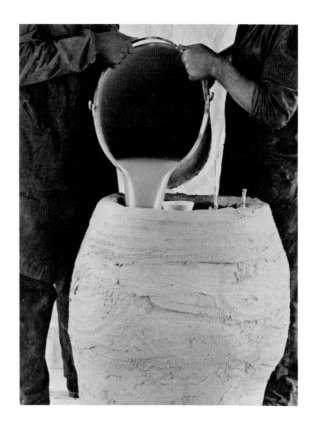

77. The thinned fireclay mixture is poured in to fill the space between the dental investment and the retaining wall to complete the entire outer fireclay mold.

78. A close-up of the pouring. Here note the funnel and the two vents.

79. The top of the finished mold. The plastic funnel has been withdrawn after having formed the pouring cup or *bevera* of fireclay. The two vent tubes have been cut off level. The fireclay surrounding their mouths is cleared away. Thus, the molten bronze issuing forth may be readily seen, signaling that the mold has been completely filled.

5: THE BURN-OUT, WHERE THE WAX IS LOST

PHASE XIX: *The oven*

80. Section of the foundry where the ovens are built. The movable crane is used for hoisting the heavier molds which cannot be handled by hand, and to lift the crucible of molten bronze when we are ready for the pour.

81. The walls of one of the ovens, made of fire-resistant brick. The fire floor is built up, and one of the fireclay molds is in place. We rebuild an oven for each new series of molds. The size is determined by the number of molds to be baked out. In most commercial foundries this process is no longer followed — a permanent oven is used and the pieces are put in without regard to their size in relation to the interior space of the oven. The wax is baked out (lost), usually with electric or gas heat, and it is most efficient, but the way we do it here dates back to the origins of this 6,000-year-old process. We don't do it for the love of tradition, but because it has very clear advantages. By building the oven precisely to the size of the molds, there is no waste space to heat, and consequently the fire is more even and controllable, insuring the molds against cracking from uneven heat. The end result is well worth the time and effort spent in custom-building an oven for each new series of molds.

82. The largest mold is selected as the "key mold" and a chimney is placed over its funnel. This chimney will permit the artisan to determine when all the wax and other burnable materials have been baked out. This is the point where the wax is "lost," thus, the term "lost wax" casting. The artisan periodically tests the chimney exhaust until all traces of greasy carbon have disappeared. For this purpose he uses a mirror or a piece of highly polished metal. When the mirror remains absolutely clean, he knows that the burn-out is complete and we can start taking the oven apart and preparing the molds for pouring.

83. The funnels of the other molds are sealed off with a sheet of metal or asbestos.

84. The sheet is weighted down and held in place by fireclay. Thus the heat coming up through the molds cannot bounce the sheet of metal out of place and allow dirty smoke to backfire down the funnel mouth, fouling the mold's interior.

85. Here the tops of all the molds, except the one with the chimney, are completely sealed. The face of the oven is being prepared. Two metal lintels support the bricks over the fire doors.

86. Inverted T-beam rafters hold the roof bricks.

87. The roof bricks are placed. Note the chimney protruding from the oven roof.

88. The entire surface of the finished oven is stuccoed with a coat of fireclay.

89. A wood fire is seen burning through both fire doors. Because it has just been started, it is going a bit too briskly, but it will burn more slowly, continuing evenly and steadily twenty-four hours a day. Depending upon the complexity and size of the molds, the burn-out takes from twenty-four to thirty-six hours and, on rare occasions, as long as a week. The fire is tended twenty-four hours a day when wood is used, as shown here. You may remember that in Cellini's book he was always running to the foundry to see how the fires were doing and whether the men had gotten drunk and were "larking about." They still do it to this day. When I'm burning out a big mold such as my first "Stampede" bronze, a real *tour de force* as far as casting goes, I stay up as much as the men do, because they have an old trick — piling on a hell of a lot of wood so they know the fire will keep burning. Then they go down to the corner and have a *fiasco* of *vino* or else just go to sleep. Meanwhile, the roaring fires have drastically changed the temperature of the oven, resulting in boiled wax and cracked molds.

It is of the utmost importance that in melting out the wax your fire be so tempered that the wax does not boil in the mould, but comes out with the greatest patience. . . . THE TREATISES OF BENVENUTO CELLINI ON GOLDSMITHING AND SCULPTURE.

PHASE XX: *Opening the oven*

90. After the burn-out has been completed, the oven is disassembled.

91. Here the artisan is lifting off one of the metal covers. Notice how the fire has marked the outer molds. They are dark at the bottom with the carbon of the burning wood, and some of the outer surface has flaked off lightly and left clean bare spots.

6: PREPARATION OF THE BURNED-OUT MOLDS AND POURING OF THE MOLTEN BRONZE

PHASE XXI: *Reinforcing the burned-out molds*

92. The first step is to spread some very dry compact sand on the floor of the casting area to create a bronze-tight footing for the molds.

93. One of the burned-out molds is set on this sand footing.

94. All of the molds have been placed on the sand bed, and the pouring cups (funnels) have again been covered with protective sheets. They are weighted down so that no extraneous dirt can fall through the pour hole into the circulatory system to block the air escape and bronze flow. Any blockage is dangerous and might cause an explosion. In addition to tube blockage, another hazard at this point is moisture — if the molds or the sand around them are damp, the whole thing could blow up like a bomb, spraying molten metal everywhere. Therefore, extreme caution must be exercised during this phase, as carelessness has caused foundry disasters which have injured or killed many men.

Not to be omitted are divers and terrible mishaps that occur from time to time, and often bring to nought all the poor master's pains. So 'tis a wise thing to profit in good time by the experience of others.
... THE TREATISES OF BENVENUTO CELLINI ON GOLDSMITHING AND SCULPTURE.

95. A portable sheet metal containing wall is placed around the molds. Additional sand is added and tamped down with a bar. It is crucial that the sand be tamped evenly. This is an art rather than a mechanically learned process. The artisan has to "feel" and take into account endless elements — the moisture of the day, the coldness of the day, the warmth of the molds themselves, the particular quality of the sand and how it packs. It must be packed just firmly enough, yet not too firmly. If it's too loose, it will not provide sufficient backup pressure to keep the mold from cracking during the pouring of the hot metal. If packed too tightly, the mold can be crushed. This has ruined castings time and again.

96. The forms are all in place in their sand bed. Note that the temporary protective coverings have been removed, and the pouring cups opened for the bronze pour.

PHASE XXII: *Founding and pouring the bronze*

97. The general area where the bronze is melted down in the furnace. Here the fire from the blast furnace is roaring up around the crucible set in the floor. To the immediate left of the fire are the hand carrier and hand pourer. In the foreground are the two-man lifting tongs, like gigantic ice tongs, used to lift the crucible out of the furnace.

98. A pig of bronze is placed into the crucible with blacksmith tongs. The pile of coke for the fire is piled up around the furnace.

99. A rod is used to clean off the extraneous materials that form slag on top, much like skim on milk.

100. Using the tongs to lift the crucible.

101. The tongs are removed from the crucible after placing it in the cradle or basket of the hand carrier.

102. The red-hot crucible and the carrier basket are hooked on the movable crane, next to the retaining wall in the pouring area. Though above ground, this area is called the "pouring pit" because its history goes back to the time when a pit for holding the molds was dug in the ground. They had not yet invented a movable wall. For extra large molds a real pit is still occasionally used. A few months ago when we poured my big "Stampede," we felt that the pressure would be so great we could not trust the metal wall, so we dug a pit in the ground and built our burn-out oven in the pit. Thus, the piece would not have to be moved for the bronze pouring. After the wax had all burned out, we simply packed sand around the mold in the same pit and then used it as a pouring pit.

103. The crane has hoisted the crucible up in place and the pouring begins.

104. Close-up of the pour showing the many molds packed in the sand inside the retaining wall.

105. Color is necessary to show the true drama of the pour. The glowing overflow from the vent holes in the mold below the crucible signifies the completion of the pour. Unless dirt or other foreign matter has fouled the mold, we can now be certain that every cranny and crevice has been perfectly filled, reproducing the final critical surface in bronze.
By the way, we do not open the molds for at least twenty-four hours after the pour. That would be a good average. It depends upon whether it's a cold or hot day. If it's in the middle of the winter, we might wait a little longer because the hot bronze exposed to the cold air would crack like splintered crystal. Bronze has — I don't understand why — quite the opposite reaction to heat than one usually associates with metal. When bronze is heated, it tends to become more brittle, and when gradually cooled becomes more malleable, as opposed to iron or steel, which become malleable and elastic at a high temperature and less so at a lower temperature. This is, I think, one of the many interesting anomalies of bronze.

7: CLEANING AND ROUGH FINISHING THE BRONZE

PHASE XXIII: *Breaking the fireclay mold*

106. The molds have been moved to the investment storage-grinding area where the fireclay is carefully knocked away. All of this fireclay will be reground and used as an additive for future investment. clay. This reused material is called "grog."

107. Half of the outer mold has been chipped away and we see the main sprues and some of the capillary runners covered with fireclay, but they are now bronze rather than cane and wax. The pouring cup, now solid bronze, is dark from dirt and extraneous matter which rose to the top. One of its purposes was to hold this matter so that it would not go into the casting.

108. Here the bronze has almost been freed from the mold. At this point the outer clay is removed with a wedge and hammer made of wood, hard enough to cut the fireclay but not hard enough to damage the bronze.

109. This close-up, aside from its pure aesthetic qualities, shows very clearly in a capsulized way the whole process, from the breather tube extension coming out of the forehead of the horse to the waste trap below his chin, to the pins of brass that still stick out here and there. Also visible are the delicate capillary runners and risers that touch and connect the critical surface of the sprues and vents. Note the one little capillary vent that touches the horse's ear. The white material (dental investment) is still clinging to the neck and back of the head. In contrast, the eye and front of the horse's head have been brushed and cleaned off, and you can begin to see the finished bronze emerge.

PHASE XXIV: *Removing the gates*

110. The casting has been completely cleaned of the outer investment, and we see now the work in its rough bronze state with most of the circulatory system still attached. It is a perfect bronze reproduction of the fully gated wax casting that we saw in Photo No. 70.

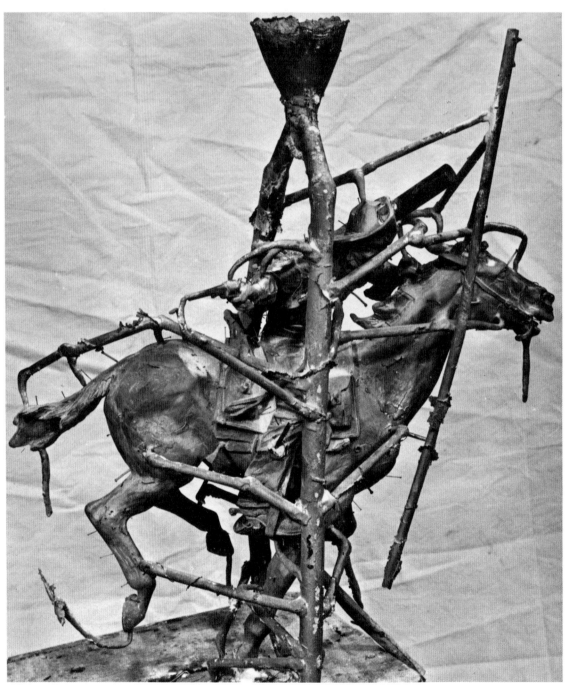

111. An artisan is cutting off the runners and vents of the circulatory system with a pair of bolt cutters. He cuts them off safely away from the critical surface. What remains can later be chased by hand with small chisels.

112. Here he lifts off a whole segment of the system.

PHASE XXV: *Emptying the bronze*

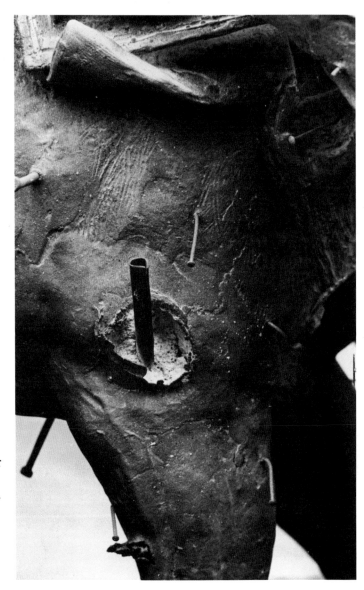

113. The auxiliary breather tube in the leg of the horse, very similar to the one we saw being placed in in the leg of the rider, Photo No. 64. The hole is there because it was cut into the wax casting. The breather tube issues forth from the hole and the core material can be seen within.

114 and 115. In these two photos, the holes are used to remove the core material. In some of the spots it's a bit difficult to get it out; thus, some may remain inside the hollow bronzes forever.

116. The removal of the breather tube extension, laying bare the tip of the main breather tube, which we saw being placed down inside the wax form before we poured the main mass of the liquid fireclay into the core.

117. The tip of the breather tube we saw being rolled by hand.

118. The necessary hole in the bottom of the horse's belly. This is left from the large support made of fireclay which we've seen from the beginning, when it supported and held the wax casting rigidly in place. Here this hole is being used to extract the bulk of the core material.

119. After the breather tube and core material have been removed, the hollow interior of the bronze is clearly visible.

PHASE XXVI: *Removing the tie rods*

120. The nails and pins being removed. These nails and pins should always be of a material related to bronze. We use brass pins since brass is closely related to bronze. In early English the words bronze and brass were quite interchangeable. Many of the statues that we know of as bronze actually are closer to brass because of their high zinc percentage, and some of the things we know of as brass are, in a technical sense, bronze because of the high tin content. We use this related material for the pins so when they fuse with the molten metal they are completely marriageable and don't have another element in them that would cause electrolysis and the ultimate decay of the bronze. Some of most famous foundries in the world use steel wire nails and iron nails, but this is not the best practice because some of the iron may oxidize in the heat of the pour and amalgamate with the bronze where it will cause ultimate damage and finally "bronze disease."

121 and 122. The extraction of one of the nails. The nails are so bulky that they very seldom fuse with the pour. The pins, on the other hand, fuse and are then simply cut off and chased.

8: CHASING AND FINAL FINISHING OF DETAILS

PHASE XXVII: *Brazing*

123. This photograph is a sort of set piece to show the bronze finishing room in its entirety. Here the rough bronze is brought, gates are cut off, and chasing and patina work are done. In this particular phase, we'll show the brazing and filling of a hole. The brazing or welding rod used for this process is poured from the same ingots that go to make up the bronze casting.

124. The hole which is going to be filled is one of those breather tube holes, a secondary one, which was used to put a breather tube and additional core material into the left front leg of the horse. The piece is held in a vise with lead covers on the jaws so that the bronze can be gripped tightly without being damaged.

125. The bronze rod used for brazing (welding). The ball of borax on the tip of the rod is used as the welding flux. The artisan first heats the entire area around the hole to be closed.

126. Heating the tip of the welding rod to the proper temperature.

127. The hole is now completely closed with the brazing material, but still rough. The dark spots around the newly closed area are bits of borax that have melted, then crystallized, and formed themselves into these little spots on the surface. They have to be thoroughly removed; otherwise, they will form a never-ending growth on the surface of the bronze which will keep emitting a white or bluish-white powder. For this reason, all welding flux must be removed before the bronze can ever be patinaed or painted successfully.

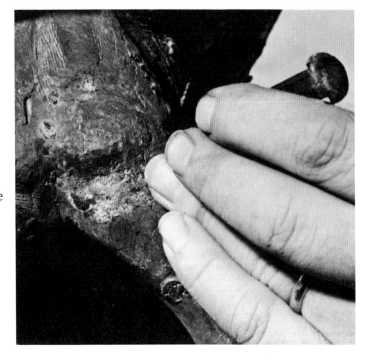

128. A chasing tool is held in one hand, slightly cushioned by the third or ring finger. The artisan, lightly tapping the tool with a hammer, shapes the metal to duplicate the original sculpture. These chasing tools are made of hard steel. We make our own in order to control the kind of cutting edge or peening surface required for each special task.

129. The artisan does rough chasing on the surface with a pneumatic drill.

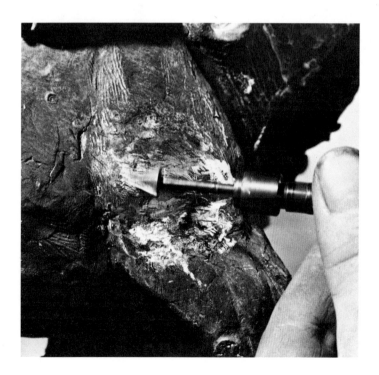

130. A close-up of the drill bit grinding.

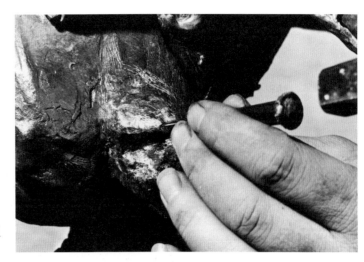

131. With one of the hand chasing tools the artisan gives the final touches to the texture on the newly brazed surface so that it will blend with surrounding textures.

PHASE XXVIII: *Placing a cold patch*

There only remains to us now to teach the method of grafting a piece into the figure should it have a defect. . . . In this case let the artificer entirely remove the defective part of the cast and make a square hole in its place, cutting it out under the carpenter's square, then let him adjust a piece of metal prepared for that spot . . . and when fitted exactly in the square hole let him strike it with the hammer to send it home, and with files and tools make it even and thoroughly finished. . . . VASARI'S COMMENTS ON THE TECHNIQUE OF CASTING BRONZE

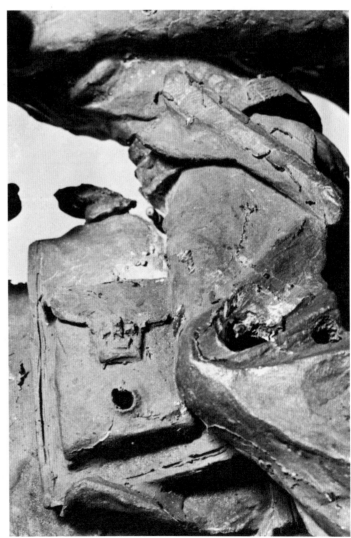

132. First the hole is cleared of all foreign matter.

133. The area surrounding the hole is countersunk to a square recess.

134. A good close-up of the prepared hole. You see that the walls of the countersunk recess are slightly undercut.

135. The hole will be filled with a bronze patch *Tassello,* which is cut from the same batch of metal used in the full casting. This bronze patch cut to the proper size fits snugly, a bit thicker than the depth of the square recess. That gives us the necessary extra material to pound down and lock into the undercut of the recess.

136. The bronze patch has been pounded and locked into place.

137. The chasing tool is used to imprint the final texture necessary to blend the surface.

138. The metal patch is now completed and indiscernible. *Tasselli* are still used in this age of modern welding methods by conscientious foundries, when welding might warp or otherwise injure the bronze. For instance, when working on a very large work or a particular surface which is unusually delicate and positioned in such a way that heat cannot be evenly applied, it is preferable to work cold using a *tassello*. In this method you do not use heat in filling a hole, or fixing up a fault or flaw in the bronze surface. Therefore, this last phase is not just put in for historical purposes — it's seen in every good foundry to this day, and where you don't find it, you don't find thoroughly trained and conscientious artisans.

PHASE XXIX: *Bronze fins*

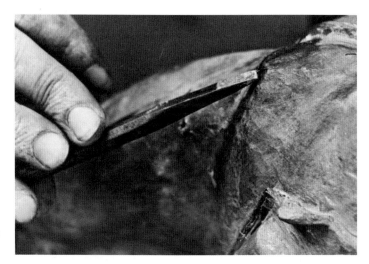

139. Here is a very clear example of another fault common in bronze casting — a fin. This is the result of slight cracks in the negative critical surface of the fireclay form. These small cracks often appear. If they are not less than hair-thin, the bronze will flow into them and produce these fins on the surface of the final piece. Here they are removed with a very sharp chisel.

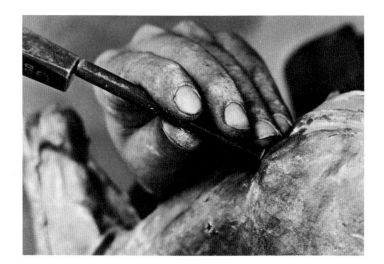

140. After the fins have been removed, the surface is cleaned with a matting tool.

These casts being finished, the workman then, with suitable tools, that is, with burins, burnishers, chasing tools, punches, chisels, and files, removes material where needed. . . . VASARI'S COMMENTS ON THE TECHNIQUE OF CASTING BRONZE

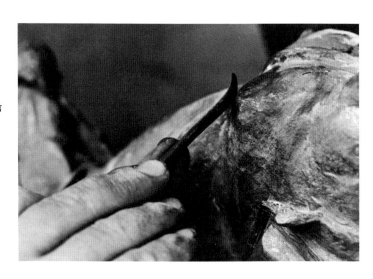

141. Finally, the surface is finished with a very fine rasp.

PHASE XXX: *Application of non-cast metal details*

142. A piece of brass wire with high copper content is pounded with a ball peen hammer until it takes on a sculptural quality with facets to capture the light.

143. A drill is used to puncture the bottom of the shank of the curb bit at the place which will receive the forward end of the bridle rein.

144. The flattened, shaped wire has now taken on the easy form of a rein. It has been introduced into the bottom ring of the bit shank and is secured with a pair of pliers.

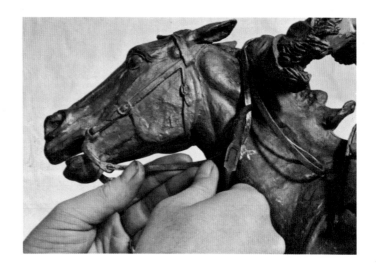

145. It is shaped along its full length to match the original sculpture and then run through the fingers of the rider's gloved hand. Silver solder is used to permanently fix the bridle rein in the bottom ring of the bit shank.

Gentle reader, let that suffice about furnaces and bronze casting, and let us now turn to other branches of the art.
THE TREATISES OF BENVENUTO CELLINI ON GOLDSMITHING AND SCULPTURE

146. The bronze work on the casting is now completed and ready for the final coloration.

THE PATINATION AND THE PAINTING OF BRONZES

MOST OF US have never known that all of the ancient cultures painted or in other ways altered the visual effects of their sculpture. They used paint or gold leaf or other precious metals in order to further refine all their sculptures, whether made of bronze, marble, terra-cotta, or wood. André Malraux documents this most clearly in *The Voices of Silence:* "With the partial exception of Egypt, the role of color in the great cultures of the past (no less distinctive and legitimate than the part played by forms) is conveyed to us by a few fragments only. . . . Greek statues were polychrome. . . . In the East statues were painted; notably those of Central Asia, India, China, and Japan. Romanesque statues were painted, so were most Gothic statues. So, it seems, were pre-Columbian idols; so were the Mayan bas-reliefs. Yet the whole past has reached us colorless."

It was the obvious and natural way for these artists to carry a statue to completion. This practice held equally for the highly evolved realism of Athens' Golden Age and the most primitive totemic sculptures. Can one truly visualize a North American Indian totem pole without its vivid, symbolic color scheme and call it complete?

This bronze which is red when it is worked assumes through time by a natural change a colour that draws towards black. Some turn it black with oil, others with vinegar make it green, and others with varnish give it the colour of black, so that every one makes it come as he likes best. . . .
GIORGIO VASARI (1511–1574)

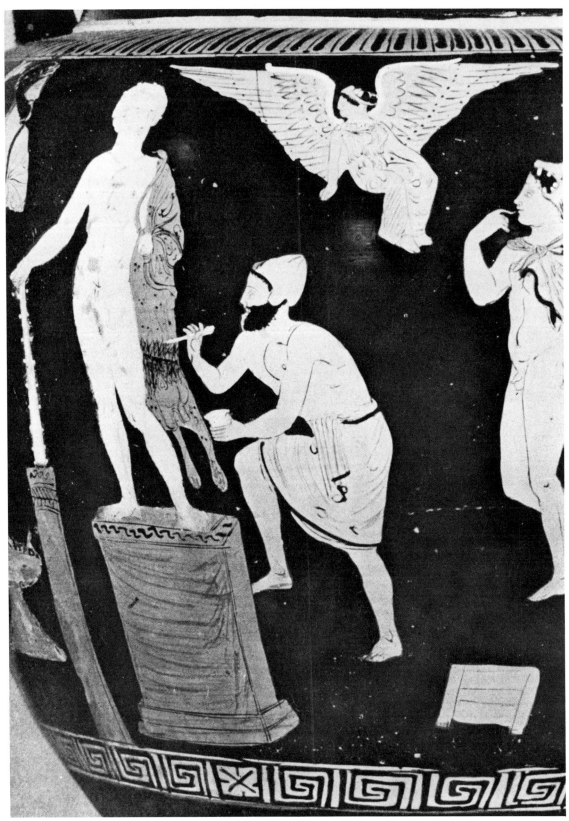

This Greek vase made in Southern Italy in the early fourth century B.C. depicts an artist painting a bronze of Hercules. Here he is putting in the detail of the lion skin. Metropolitan Museum of Art, Rogers Fund, 1950.

The practice of painting and decorating sculpture fell into disuse and was lost sight of during the barbarian invasions that overwhelmed Rome. The fine arts in general fell into decay, and the works were defaced, destroyed, or torn down and forgotten. Many marble and bronze statues were lost and buried in rubble or at sea, where they remained until the re-dawn of Humanism in the Renaissance.

Over 1200 years had passed, and the slow work of the acids and the attrition of earth, water, and air had decomposed all of the paint and caused a wild variety of effects on the exposed bronze surfaces through corrosion and oxidation. This natural type of surface, called a *Patina Nobile* (noble patina), is formed from long exposure to the air, rain, and dust. The high copper content of bronze (from 65 to 90 per cent) will oxidize and create this protective skin of carbonate of copper. However, when bronzes were haphazardly buried in many different kinds of soil or water and exposed to unknown combinations of chemicals, the most varied and sometimes startlingly beautiful effects were formed by a combination of corrosion and encrustation.

A perfect example of early Renaissance sculpture showing the effect of long exposure to the elements. Donatello's GATTAMELATA *in the Piazza del Santo, Padua, Italy.*

Newly discovered during the Renaissance, these bronzes were accepted with unquestioning reverence, as divine legacies from Ancient Athens, and provided inspiration for an entire age. The colorations of this natural oxidation, these marvelous, but meaningless, accidental patinas were imitated by the artists of the fifteenth and sixteenth centuries. They taught themselves to duplicate these stunning effects, wrought in the past only by unthinking chance, and created the art of patination. They could simulate in a few hours or weeks the patinas that unaided nature had required more than one thousand years to produce. As we can gather from Vasari's quote, they used any means possible.

Accidents often produce great innovations and the patina is an outstanding example. Its relatively uniform coloration gives us an entirely fresh sense of a sculpture's overall form, mass and movement. These important aspects are far less apparent when the work has been refined with local colors or decorative patterns. Therefore, there are definite advantages to each of these opposed methods.

I for one like to create both patinaed and realistically painted versions of many of my works. A striking case in point is the very sculpture chosen to illustrate this book.

The method I mainly employ for forming a patina is described in the following section. The effect can be noted here on Rodin's small bronze study for his great monument, "The Burghers of Calais."

Despite this practice of falsely aging bronzes, born in the Renaissance, many sculptors have continued to paint and decorate their sculptures right up to the present time. The gilding of bronzes, dating back to ancient Egypt, has continued to resurface in every age. Outstanding examples are Donatello's (1386–1466) Saint Louis as a young bishop, and twentieth century works, such as Anna Hyatt Huntington's monumental equestrian statue of Jeanne d'Arc in Paris and Paul Manship's series, "The Labors of Hercules." Pliny, the Elder (A.D. 23–79) discusses the practice at some length: "When also things made of copper [bronze] are gilded, a coat of quicksilver is applied underneath the gold leaf and keeps it in its place with the greatest tenacity."

He also speaks of "a method discovered in the Gallic provinces to plate bronze articles with white lead [tin] so as to make them almost indistinguishable from silver."

An excellent example of still existing painted sculpture is the entire façade of the great Gothic cathedral in Freiburg, Germany (ca. 1200). Its façade, encrusted with innumerable figures, and architectural elements such as columns, capitals, and moldings — all are painted in most vivid colors. As for painting bronzes, the famous Mary Magdalen by Donatello in the Baptistery of the Duomo in Florence, Italy, was so completely coated with dirt and grime that everyone assumed that it had purposely

This small bronze study by Auguste Rodin, 1840–1917, for his great monument, THE BURGHERS OF CALAIS, *illustrates the traditional patina with its uniform coloration. The author's collection.*

been cloaked in a dark uniform patina by the sculptor. And it was not until the disastrous flood of November 1966 had washed away some of this coat of dirt, that the experts realized the Magdalen had been painted realistically. It is a life-size figure and has now been completely restored, showing the natural flesh and hair tones.

In the Museum of Fine Arts in Marseilles, France, there is one of the finest collections of vividly painted small figures and portrait busts by the French caricaturist Honoré Daumier (1808-1879). Picasso, too, has painted many of his sculptures, as has Giaccometti. But to bring the argument closer to the subject of my sculpture, I have reproduced a small head of a longhorn steer by Charlie Russell. He painted all of the originals that he made in wax or wood or plaster because he wanted us to know just what color a particular steer or horse was or what characteristic colors his Indians and cowboys wore. For him, just as for the ancient Greeks, the American Indians, or myself, a work of sculpture was not complete until he had painted on the local colors appropriate to the sculptured forms. When Russell's colored originals were sent from Montana to the Roman Bronze Foundry in New York, the molds were made and many bronzes cast of each work. Just who made the decision to change Russell's original artistic statement by finishing all of the bronze castings with a uniform patina, I don't know, but it seems self-evident that at least a part of the bronze castings of each subject should have reproduced the original color as well as the form. Therefore, I believe these bronzes have come down to us incomplete.

The accidents of time and history have given us the uniform coloration of the patina. It has become so deeply ingrained in our tastes that it is definitely here to stay, and it has many sound arguments on its behalf. In order to present a complete picture, I have chosen to illustrate both methods.

Head of longhorn steer by Charles Russell, 1864–1926. He always painted his original models in realistic colors before sending them to the foundry for casting. The final bronzes, however, were finished with a uniform patina. The author's collection.

PATINATION

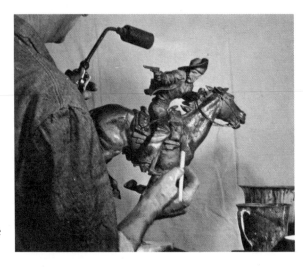

FIG. 1. The acids for the patina in this example are applied in a liquid state by brush when the bronze surface has been heated with a blow torch. The various acids stand to the right, handy to the craftsman's brush as he heats the bronze and applies the liquid, time after time until the desired effect is achieved.

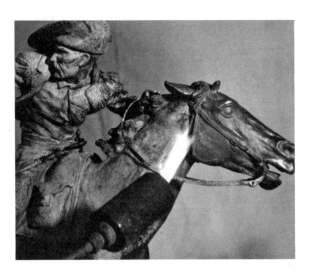

FIG. 2. A close-up of the torch flame and the green coloration as it begins to cover the bronze. This green is made by first dissolving bits of pure copper scrap in a bath of nitric acid. When the copper has completely disintegrated, one part of the resulting solution is adulterated with at least two parts water. The pores of the bronze surface open when heated so that the various acids will penetrate and impregnate the skin of the metal. When the metal cools and the pores close, the acids become locked in and form an integral part of the surface. It is not just a film laid on top, a chemical change has been started.

FIG. 3. After the entire work has been treated with this green acid (copper nitrate), the same procedure is repeated and a coating of brown acid (ferric nitrate) is applied.

FIG. 4. We see the head and forequarters of the horse and most of the rider treated with this second application, while the hindquarters, tail, and the pistol arm and pistol still clearly show the first green reaction.

FIG. 5. The entire work now has the final tone desired, and while the bronze is still warm a thin coating of beeswax is brushed on and sinks into the surface. The bronze must be left to cool and set for several days in an extremely dry place. After this it is rubbed briskly with a soft wool flannel cloth, and the work is complete.

PAINTING

My bronzes are painted with a tempera paint. True tempera is an emulsion of drying oils and water. My own is one of many closely related versions of a basic formula reported as early as the fourteenth century. It is safe to surmise that similar combinations were used in pre-Christian Greece and Rome. It is called Tempera *Grassa* (Fat Tempera). My formula is:

> 2 parts whole egg
> 1 part Damar Resin Varnish (5-lb. cut)
> 1 part boiled linseed oil
> 2 to 4 parts 7 per cent glue water (finest rabbit skin glue 7 per cent by volume dissolved in warm distilled water)

FIG. 1. Dry white lead powder is mixed into a very stiff paste with the tempera emulsion.

FIG. 2. After this it is further ground with a marble hand muller on a slightly abrasive marble slab until each particle of dry powder (pigment) is completely dispersed and surrounded by the liquid emulsion.

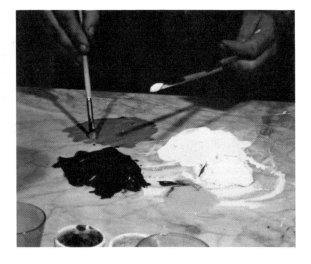

FIG. 3. This procedure is repeated for each of the colors. In the next step, not illustrated, the bronze surface is given a thin first coat of the best grade of red iron oxide automobile primer-sealer to isolate and keep the bronze from any unfavorable chemical reactions that might be caused by the eggs and oils in the emulsion. In the past a pure glue water sizing was used, followed by a primer coat of white gesso or red bolus earth. The modern automobile primer-sealer is used because it forms a more resistant and elastic film. The freshly ground tempera paint is applied directly over this surface.

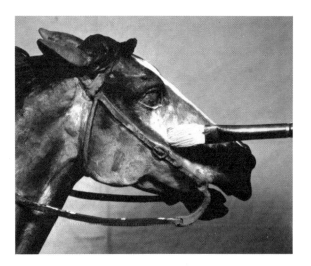

FIG. 4. This first tempera coat is always a general tone called the "lay-in" or "under-painting."

FIG. 5. When this has dried I proceed with combinations of dense-covering (opaque), half-covering (translucent), and veil-thin (transparent) glazes until the desired effect is achieved. When the paint is thoroughly dry, it is varnished with a mat (non-glossy) varnish. When this is dry, it is finished with a thin coat of wax warmed in a double boiler. When the wax is dry the areas which should have a luster are polished, in this instance, the gun metal and the characteristically shiny areas on the leather of the saddle, leggings, etc. The materials like flannel shirts and saddle-blankets can be left mat as they are in reality.

GLOSSARY OF TECHNICAL TERMS

BAKE-OUT. The process in which the reproductory wax is "lost." *See* BURN-OUT.

BASKET. *See* SHANK.

BEVERA. *See* POURING CUP.

BRASS. A metal alloy consisting of 60 to 90 per cent copper with the balance of zinc. Small quantities of nickel, tin, and lead may also be present.

BRAZEN. All metal alloys in which the basic element is copper.

BRAZING. A method of welding (joining) together two pieces of brazen metal. The metal of the casting is fused with a brazing rod of the identical material (photos 123 through 131).

BREATHER TUBE. A tube within the fireclay core or interior investment of the reproductory wax shell which permits the escape of gases formed within the mold during the casting of the bronze. In this case the hollow tube is open at both ends and along its entire length (photos 32 through 35, 66 through 70, 113, 116, 117).

BRONZE. A metal alloy consisting of 60 to 90 per cent copper with the remaining portion of tin. It can also contain nickel, zinc, lead, or silver in smaller proportions. In Biblical writings and up to the 19th century, brass and bronze were interchangeable. Many statues known as bronze are, technically speaking, brass because of the high zinc content.

BRONZE DISEASE. A slow deterioration of the bronze. The cause of this condition has been much discussed, but no definite conclusions have been reached. In general it seems to be caused by an electrolytic reaction of the bronze with unrelated metals such as iron and steel. Thus it seems advisable to use only brazen metal in reinforcing and finishing a work (photo 120).

BRONZE PATCH (Italian — *tassello*). A cold method of filling holes by locking a piece of metal (same alloy as the casting) into the hole. It is used when the hot method (brazing) is difficult or dangerous to the casting (photos 132 through 138).

BURN-OUT. The process in which the reproductory wax is "lost." The molds are placed in an oven or kiln where sustained heat completely consumes the burnable material in the mold. Thus the reproductory wax is burned out. Burnable material in the circulatory system is also consumed so that the channels in the fireclay mold remain clean, allowing the molten metal to flow and the gases to escape. At the same time, the sawdust burns out of the core, leaving it porous. The wax, having burned out, leaves the so-called cavity or pattern of itself between its fireclay core and the outer fireclay investment or mold (photos 80 through 91).

CHASING. The cleaning up of seams, and the repair of surface imperfections. It is also the careful art by which detail and sharpness of texture are restored or altered (photos 128 through 131, 139 through 141).

CHOKE. A reduced section, usually in the runner system, which restricts the flow of metal into the cavity. The purpose of the choke is to prevent dross, oxides, and so forth from getting into the casting cavity. The choke causes the metal to be restricted, the sprue then fills up, and the pouring is done with the pressure of a full sprue until the entire mold is filled (*see* schematic drawing on opposite page).

CIRCULATORY SYSTEM. The network of large and small arteries through which the molten metal is poured into the mold. It also allows air and gases to escape during casting (photos 66 through 70, 109 through 112, and schematic drawing).

CIRE PERDUE. French term used universally for the lost wax method of casting (*see* page 4).

COLLATE. Italian term for circulatory system.

CORE. *See* INVESTMENT.

CORE ANCHORS. Metal stays, preferably brazen pins and nails, act as tie rods and hold the interior and exterior investments in place after the reproductory wax has been burned out (photos 59 through 61, 120 through 122).

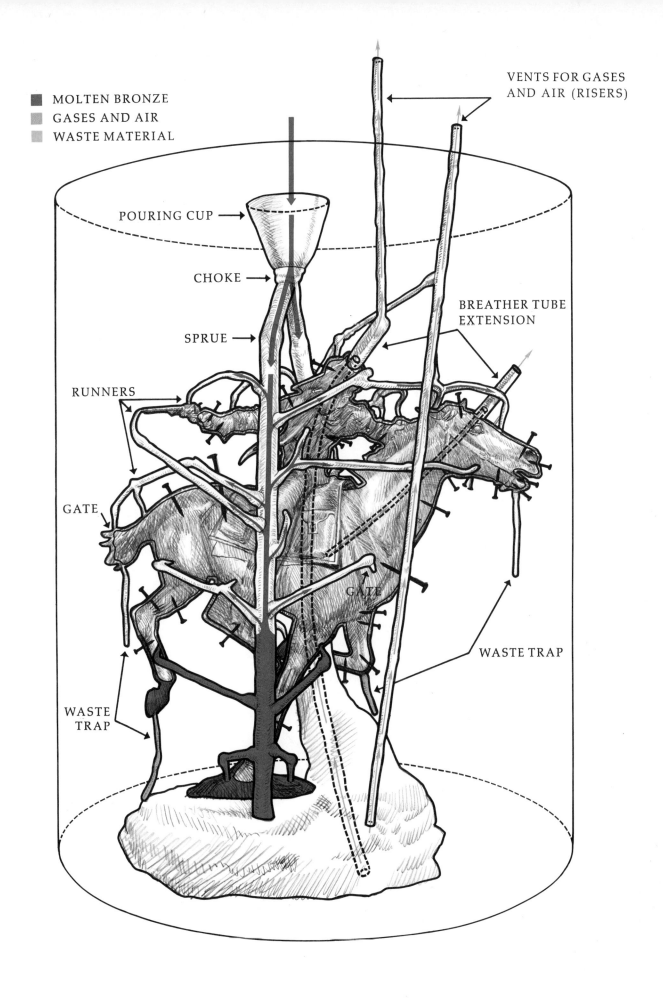

CORE VENT. *See* BREATHER TUBE.

COUCHE. Also referred to as the plaster outer shell or mother mold, this is the rigid plaster shell which houses and protects the flexible rubber (or other elastic substance), keeping it rigid while the hollow reproductory wax shell and its inner core of fireclay are built up inside (*see* PHASE IV of text and photos 18 through 27).

CRITICAL SURFACES—POSITIVE, NEGATIVE. Begins with the surface of the original as created by the artist. This surface must be reproduced exactly in negative and positive impressions through the various stages of the reproductory process. Thus:
1. Original — positive critical surface, permament (photo 15)
2. Master mold — negative critical surface, permanent (photo 16)
3. Reproductory wax — positive critical surface, temporary (photos 55, 56)
4. Outer fireclay dental investment — negative critical surface, temporary (photos 71 through 74)
5. Cast bronze — positive critical surface, final permanent surface (photo 146)

This is the principle of the reproductory process, but it is evident that the artist does not resign himself to increasingly inexact impressions of his original. When he is presented with a positive reproduction of his original in reproductory wax, and later in bronze, he freshens this surface and alters it as he chooses. Thereby each bronze cast differs, and becomes a truly original and inimitable casting.

CRUCIBLE. A pot of refractory material used for founding metals (photos 97 through 105).

FINS. *See* FLASHING.

FIREBRICK. *See* FIRECLAY.

FIRECLAY. Refractory clay capable of withstanding high temperatures (photos 36, 71 through 79).

FIRECLAY MOLD. *See* INVESTMENT.

FLASHING. Casting faults to be rectified. They occur in the reproductory wax where the wax seeps through the joints of the various parting lines of the rubber mold or on the surface of the bronze where the fireclay mold has cracked slightly during the burn-out or from the pressure of the molten bronze and its gases (photos 51, 139 through 141).

FLUX. A substance to remove impurities from molten metal. When the bronze is heated in the crucible for the pour, flux may be added (here aluminum is used). It should react at melting temperature, remove impurities, and be easily separable from the melt itself so as not to run into the mold. The product formed by reaction with the flux is called slag. Flux is also used in welding or brazing. In this case its purpose is to remove the layer of oxidation formed on the cast bronze around the area to be brazed. With the oxidation removed, this area of the casting will melt at the same temperature as the brazing rod and fusion is achieved (photos 125 through 127).

FOUNDING. Melting and pouring of metals (photos 97 through 105).

FUNNEL. *See* POURING CUP.

GATE. The point of contact between the runner and the cavity (photos 66 through 70 and schematic drawing).

GATING. Vernacular term for entire structure of circulatory system (photos 66 through 70 and schematic drawing).

GROG. The reground investment material made of plaster and firebrick. Each fresh batch of investment consists of from one-third to one-half grog (photo 106).

HOLLOW WAX POSITIVE. *See* REPRODUCTORY WAX.

IN-GATE. *See* GATE.

INVESTMENT:

Core investment. Interior portion of the fireclay mold (photos 28, 36 through 39, 113 through 119).

Outer investment. The negative critical surface and its robust casing, created by covering the positive critical surface of the finished reproductory wax. The first layer of this mold may be formed of fine dental investment painted on the reproductory wax. It is then surrounded by a rougher fireclay. These molds are not reusable but are destroyed after each casting and ground up so that the fireclay or grog itself can be reemployed. This investment, both inner and outer, is made anew for each bronze to be cast (photos 71 through 79).

KEY MOLD. The largest form to be baked out in a particular oven. The chimney is attached to this mold because it is the largest and takes the longest time to burn out. It most accurately insures that the organic material has been burned out of all the molds. Thus, it is the safest indicator of the state of the molds. The vapors which pass through its chimney will indicate when the burn-out is completed (photo 82).

MASTER MOLD. The permanent critical negative impression of the positive critical surface of the original wax from which positive reproductory wax shells are made. It can be of various substances — interlocking plaster piece mold, gelatin, moulage, glue, or cold-setting synthetic rubber as used here (photo 16).

MAT or MATT, also MATTE. A rough or granular surface without gloss (Fig. 5, page 111).

MOLD. See MASTER MOLD, INVESTMENT.

MOLD CAVITY. The hollow between the core and the outer investment created by the lost wax, and into which the molten metal is poured.

ORIGINAL. The artist's actual creation in any material which will then be reproduced in bronze (photo 15).

PATINA. In the broadest sense, it is the surface appearance or coloring of the bronze. The term can be applied to at least three categories:

Patina falsa (false patina). Color added to the surface of the bronze without making a chemical bond with that surface. This can be oil paint, varnish, colored gesso, or plaster.

Patina chemical intentional. A colored encrustation and corrosion of the bronze surface intentionally produced by chemical agents, such as sulphides, oxides, and salts. An "artificial" patina simulates the effects of the natural one created by exposure to the elements and this can be achieved in a very short time (pages 108 and 109).

Patina nobile. Italian term for noble patina. Formed on any copper, bronze, or brass surface exposed to the elements for many years. It forms a protective coating of carbonate of copper. This should never be removed or disturbed as it is irreplaceable except under equal conditions and periods of time.

PATTERN. See MOLD CAVITY.

PECE GRECA. Italian term for gum resin. One of the ingredients used in making the wax alloy for the reproductory wax (photo 21).

PINS. See CORE ANCHORS.

POLYCHROME BRONZE. From the Greek poly (many) and chrome (colors). Used here to describe a bronze painted with many colors (pages 110 and 111).

POURING CUP (Italian term is *bevera*). The widened mouth in the top of the fireclay mold through which the molten bronze is poured (photo 79 and schematic drawing).

POURING PIT. The confined area in which fireclay molds are secured for receiving the molten bronze. Originally it was actually a pit in the ground. Now its protective function is largely served by a movable metal retaining wall into which the molds are placed and surrounded with supporting packed or "tamped" sand (photos 95, 96, 102).

REPRODUCTORY WAX. The hollow wax reproduction of the critical surface of the original cast in master mold (photos 50 through 58).

RESPIRATORI. Italian term for the vents.

RISERS. The beginning of the "vein" half of the circulatory system. Corresponding as opposites to the runners or "arteries," these channels lead either out of the cavity itself, or out of the runners, up and out of the vents to the top of the mold form. They allow the gases to escape during casting, and as they fill with molten metal, they provide a source of excess metal to compensate and replenish the shrinkage of the metal cooling in the critical cavity (photos 66 through 70 and schematic drawing).

RUNNER. The horizontal channel between the sprue and the cavity. Runners should incline slightly from the sprue to the cavity to increase and eliminate air bubbles (photos 66 through 70 and schematic drawing).

SEAMS. *See* FLASHING.

SHANK *or* CARRIER. The metal basket into which the crucible is placed for carrying and for controlling the pouring of the molten bronze (photos 97, 101 through 105).

SLUSHING. The crucial process which determines the thickness of the shell of the reproductory wax, and ultimately, of the bronze. Molten wax is poured into the wax-coated master mold and, while still fluid, is slushed around in the mold until sufficient wax adheres to the negative surface and has hardened. Excess wax is then poured out (photos 28 through 31).

SOLDER. Low melting point metals — lead, tin, or silver, used singly or in combination (photo 145).

SOLDERING. Joining two metal pieces by the use of molten solder.

SPRUE. The vertical channel under the pouring cup which leads the molten bronze to the mold cavity through smaller branches called runners (photos 66 through 70, 107, 108, and schematic drawing).

SUPPORTING BASE, FIRECLAY. The extension of the fireclay core upon which the reproductory wax will stand securely while it is being retouched, provided with its circulatory system, invested, and in general prepared for casting (photos 40, 41, 51, 118).

TASSELLO. *See* BRONZE PATCH.

TERRA-COTTA. A brownish orange fired clay. Literal translation is "baked earth."

PRIMITIVE GATING SYSTEM. *A drawing of one of the Amuq Plain statuettes, c. 4000 B.C., illustrating the simple gate and vent system undoubtedly used by the early primitive sculptors. Wax extensions were added to the top and bottom of the original wax model, as shown here. It was then completely enveloped in a thick coating of fine liquid clay with the wax extensions protruding slightly. This clay mold was heated in a fire until it was hard as brick and the wax figure and extensions had burned out. Melted bronze was then poured through the upper extension (the gate) of the hollow clay mold, forcing the air out through the lower extension (the vent). When the bronze had cooled, the baked clay was chipped off, leaving the bronze casting. The gate and vent extensions were removed and the small figure smoothed with ancient versions of today's chisels, rasps, and hammers.*

TRAPS. Downward projections, either directly from lower parts of the cavity or from the circulatory system. During casting these traps catch the waste material left in the cavity or pattern after the burn-out, thus preventing the material from forming faults in the bronze (photos 66 through 70, 109, and schematic drawing).

VENTS. The final part of the circulatory system. These tubes lead back up to the top of the mold from the risers, and allow the gases to escape from the cavity. When the bronze finally issues forth from these vents, the pouring is complete (photos 66 through 70, 105, and schematic drawing).

WASTE TRAPS. *See* TRAPS.

SELECTED BIBLIOGRAPHY

BOOKS

Ackerman, P. *Ritual Bronzes of Ancient China.* New York: Dryden, 1945.

Arnold, Edward. *The Treatises of Benvenuto Cellini on Goldsmithing and Sculpture.* London: Guild Press, 1898.

Ashton, L. *An Introduction to the Study of Chinese Sculpture.* London: Benn, 1924.

Bailey, K. C. *The Elder Pliny's Chapters on Chemical Subjects.* London: 1929.

Bachhofer, L. *A Short History of Chinese Art.* New York: Pantheon, 1946.

Boardman, John. *Greek Art.* New York: Frederick A. Praeger, Inc., 1964.

Brown, G. B., ed. *Vasari on Technique.* London: Dent, 1907. New York: Dover Books, 1960.

Burgess, J. *Buddhist Art in India.* London: Quaritch, 1901.

Casson, S. *The Technique of Early Greek Sculpture.* Oxford: 1933.

Cellini, Benvenuto. *The Autobiography of Benvenuto Cellini.* Translated by J. Addington Symonds. New York: P. F. Collier & Son, 1910.

Dimand, M. S. *A Handbook of Muhammadan Art.* New York: Metropolitan Museum of Art, 1944.

Enciclopedia dell' Arte Antica. Vol. 2. Rome: Istituto Treccani, 1959.

Enciclopedia Italiana. "Fusione." Vol. XVI, pp. 211–223. Milano: Istituto Treccani, 1932.

———. "Bronzo." Vol. III, pp. 928–942. Milano: Treccani-Rizzoli, 1930.

Encyclopedia of World Art. "Bronze." Vol. XV (Index). New York: McGraw-Hill, 1968.

Fink, C. G. & Koop, A. J. *The Restoration of Ancient Bronzes and other Alloys.* New York: Metropolitan Museum of Art, 1925.

Forbes, R. J. *Metallurgy in Antiquity.* Leiden, Netherlands: E. J. Brill, 1950.

———. *Studies in Ancient Technology.* Vols. 8 & 9. Leiden, Netherlands: E. J. Brill, 1964.

Franciscis, Alfonso de. *Il Museo Nazionale di Napoli.* Napoli: de Mauro, 1963.

Frankfort, Henri. *Sculpture of the Third Millennium, B.C. from Tell Asmar and Khafajah.* Chicago: The University of Chicago Oriental Institute Publications, Vol. XLIV, 1939.

———. *The Art and Architecture of the Ancient Orient.* Harmondsworth: Pelican, 1954

Geck, Francis J. *Bibliography of Italian Early Renaissance Art.* Vol. 6. Boulder, Colorado: 1932.

———. *Bibliography of Italian Late Renaissance Art.* Vol. 8. Boulder, Colorado: 1934.

Gangoly, O. C. *South Indian Bronzes.* Calcutta: Indian Society of Oriental Art, 1915.

Hoffman, Malvina. *Sculpture Inside and Out.* New York: W. W. Norton & Co., 1939.

Hsi-ch'ing Ku-chien. "Survey of Antiquities in the Hsi-ch'ing." 42 volumes, 1751.

Hsüan-ho-Po-ku-t'u-lu. "Illustrated Record of Antiquities in the Hsüan-ho." 30 volumes, A. D. 1107.

Janson, H. W. *History of Art: A Survey of the Major Visual Arts from the Dawn of History to the Present Day.* First printing. New York: Harry N. Abrams, Inc., 1962.

Kar, C. *Classical Indian Sculpture.* London: Tiranti, 1950.

Karlgren, Bernhard. *A Catalog of the Chinese Bronzes in the Alfred F. Pillsbury Collection.* Minneapolis: University of Minnesota Press, 1952.

Kluge, K. & Lehmann-Hartleben, K. *Die Antiken Grossbronzen.* 3 volumes. Berlin: de Gruyter, 1927.

Koop, A. J. *Early Chinese Bronzes.* London: Benn, 1924.

Kramkisch, Stella. *The Art of India.* London: Phaidon, 1954.

Lankheit, K. *Florentinische Barockplastik.* Munich: Bruckmann, 1962.

Levey, M. *Chemistry and Chemical Technology in Ancient Mesopotamia.* Philadelphia: 1959.

Lü Ta-Lin. *Kao-Ku-t'au.* "Illustrated Treatise on Antiquities." 10 volumes, A. D. 1092.

Malraux, André. *The Voices of Silence: Man and His Art.* Garden City, New York: Doubleday & Co., Inc., 1953.

Marinatos, S. *Creta e Micene.* Florence: Sansoni, 1960.

Mayer, Ralph. *The Artist's Handbook of Materials and Techniques.* 3rd ed., rev. New York: Viking Press, 1970.

Parlanti, E. J. *Casting a Torso in Bronze.* London: Alec Tiranti Ltd., 1953.

Planiscig, L. *Piccoli Bronzi Italiani del Rinascimento.* Milano: Treves, 1930.

Plinius Secundus, C. *Natural History.* Translated by H. Rackham, Cambridge, Mass.: Harvard University Press, 1967.

Pope-Hennessy, J. *Italian Gothic Sculpture.* London: Phaidon, 1958.

———. *Italian Renaissance Sculpture.* London: Phaidon, 1958.

Rich, Jack C. *The Materials and Methods of Sculpture.* Second printing. New York: Oxford University Press, 1956.

Richter, G. M. A. *Greek, Etruscan and Roman Bronzes.* New York: Metropolitan Museum of Art, 1915.

———. *A Handbook of Greek Art.* London: Phaidon, 1959.

Roast, Harold J. *Cast Bronze.* Cleveland, Ohio: The American Society for Metals, 1953.

Ross, E. Denison. *The Art of Egypt Through the Ages.* London: Studio, 1931.

Swarup, S. *The Arts and Crafts of India and Pakistan.* Bombay: Taraporevala, 1957.

Swedish Institute in Rome. *Etruscan Culture, Land and People.* New York: Columbia University Press and Malmö: Allhem, 1962.

Wear, Bruce. *The Bronze World of Frederic Remington.* Tulsa: Gaylord, Ltd. (Art Americana), 1966.

Westendorf, Wolfhart. *Painting, Sculpture, and Architecture of Ancient Egypt.* New York: Harry N. Abrams, Inc., 1968.

Zimmer, Heinrich. *The Art of Indian Asia.* New York: Bollingen Series XXXIX, 1955.

PERIODICALS

Barnard, N. Review of *Bronze Casting and Bronze Alloys in Ancient China*. Nagoya, Japan, 1961. *Art Bulletin*, Vol. XLV, No. 4, December 1963, pp. 394–396.

Baumgartel, E. "The Cire-Perdue Process." *Antiquity*, Vol. XXXVII, No. 148, December 1963, pp. 295–296.

Bearzi, B. "Considerazioni su la formazione delle patine e delle corrosioni sui bronzi antichi." *Studi Etruschi*, Vol. XXI, serie II, 1950–51, pp. 261–266.

———. "Considerazioni di tecnica sul S. Ludovico e la Guiditta di Donatello." *Bollettino d'Arte*, Vol. XXXVI, serie IV, 1951, pp. 119–123.

———. "Esame Technologico e Metallurgico della Statua di San Pietro." *Commentari*, anno XI, No. 1, 1960, pp. 30–32.

———. "Relazione tecnica sul restauro dei bronzi."*Bollettino d'Arte*, anno XLV, serie IV, 1960, No. I–II, January–June, pp. 42–44.

Cambi, L. "Problemi della mettalurgia Etrusca." *Studi Etruschi*, Vol. XXVII, serie II, pp. 415–432.

Cialdea, U. "Restoration of Antique Bronzes." *Mouseion*, Vol. XVI, 1931, p. 57.

Coomaraswamy, Ananda K. "Indian Bronzes." *Burlington Magazine*, Vol. XVII, No. LXXXVI, May 1910, pp. 86–94.

Farnsworth, M. "The Use of Sodium Metaphosphate in Cleaning Bronzes." *Technical Studies in the Field of Fine Arts*. Fogg Art Museum, Vol. IX, 1940, p. 21.

Ferguson, J. C. "An Examination of Chinese Bronzes." *Smithsonian Report* for 1915 from a translation by Liang T'sung-shu in 1767.

Fink, C. G. & Polushkin, E. P. "Microscopic Study of Ancient Copper and Bronzes." *Metals Technology*, Vol. III, 1936. American Institute of Mining Engineers.

———. "Microscopic Study of Ancient Bronze." *Transactions of the A.I.M.E.*, Vol. CXXII, 1936, pg. 90.

Gettens, R. J. "Mineralized Electrical Treatment and Radiographic Examination of Copper and Bronze Objects from Nuzi." *Technical Studies in the Field of Fine Arts*. Fogg Art Museum, Vol. I, 1933, p. 119.

Kluge, K. "Bronze work of the Geometric Period." *Journal of Hellenic Studies*, Vol. XLII, 1922, pp. 207–219.

Mallowan, M. E. L. "The Amuq Plain." *Antiquity*, Vol. XXXVII, No. 147, September 1963, pp. 185–192.

Matsumo, T. "Constituents of Ancient Bronze and the Constitutional Relations Between the Original Alloy and the Patina." *Journal of the Chemical Industry of Japan*, Vol. XXIV, 1921, p. 1369.

McEwan, C. W. "The Syrian Expedition of the Oriental Institute of the University of Chicago." *American Journal of Archaeology*, Vol. XLI, 1937, pp. 8–16.

Ridgway, B. S. "Stone and Metal in Greek Sculpture." *Archaeology*, Vol. XIX, No. 1, January 1966, pp. 31–42.

Stubbings, F. H. "A Bronze Founder's Hoard at Mycene." *Annual of the British School at Athens*, XLIV, 1954, p. 292.

Stucchi, Sandro. "Gruppo bronzeo di Cartoceto." *Bollettino d'Arte*, 1960, Anno XLV, serie IV, No. 1–2, January–June, pp. 7–42.

Tefft, Elden C., ed. "National Bronze Casting Conference Proceedings." Lawrence: University of Kansas, 1960, 1962, 1964, International Conference — 1966.

———. "Sculpture Casting in Mexico: A Report on Bronze Casting of Sculpture in Mexico." Lawrence: University of Kansas, 1960.

———. "Lost Wax Sculpture: Foundry Equipment, Sources, and Prices." Lawrence: University of Kansas, 1964.

ACKNOWLEDGMENTS AND CREDITS

THIS BOOK has finally seen the light of day because of the unstinting help and advice of so many people; many of them are dear friends. First, Richard Fremantle, who did extensive historical research for me; Rudy Wunderlich, president of the Kennedy Galleries, New York, who originally suggested the need of a book on this subject; my wife, Sarah Mason Jackson, for her patient, understanding cooperation during the long initial stages; the sculptors, Jacques Lipchitz, Thomas McGlynn, Bruce Moore, Henry Moore, Paul Sutton, and the late Paul Manship, for their encouragement and professional advice; Miss Peggy Haines, Mrs. Viola Packer, and my secretary Miss Angela Menconi for typing the endless revisions and sorting photographs; the writers Linda Arkin and Jerry McGuire for their criticism and aid in polishing the test. I wish to express my appreciation to Paul Weaver, the publisher of Northland Press, for his enthusiasm and for the infinite pains he and his entire staff have given to the realization of every aspect of this book, settling only for the best; to Mrs. Doris Monthan for endless help in every facet of editing and proof reading, to Robert Jacobson for his excellent design, the Italian photographer Lido for his keen grasp of the project and fine photography. And last but not least, I wish to thank all the men in my foundry for their cooperation during the production of the photographs: Mario Caniparoli, Franco Bertoni (foreman), Alfredo Domenici, Mario Giannecchini, Paolo Hauri, Roberto Marsili, Giuseppe Poli, and Ivo Ricci.

PHOTOGRAPHY:

Studio Limar, Camaiore, Lucca, Italy
Studio Sabella, Marina Di Piestrasanta, Italy
Richard Fremantle, Florence, Italy

Male Portrait Head from Ife, Nigeria, page 4, by Eliot Elisofon
Warrior on Horseback, page 4, by John Webb
Parvati, page 6, by Ferenc Berko,
courtesy of Mrs. A. K. Coomaraswamy

DESIGNED BY ROBERT JACOBSON
COMPOSED IN LINOTYPE ALDUS
WITH DISPLAY LINES IN WEISS INITIALS I
PRINTED AT NORTHLAND PRESS
ON MOUNTIE WARM WHITE
BOUND BY ROSWELL BOOKBINDING

739.512
106584
Jackson, Harry
Lost wax bronze casting.

PUBLIC LIBRARY
City of Long Branch, N. J.